30th Joint Russian-German Meeting on ECRH and Gyrotrons 2018

EPJ Web of Conferences Volume 187 (2018)

Nizhny Novgorod, Russia
17 – 24 June 2018

Editor:

V.E. Zapevalov

ISBN: 978-1-5108-7062-8

Printed from e-media with permission by:

Curran Associates, Inc.
57 Morehouse Lane
Red Hook, NY 12571

Some format issues inherent in the e-media version may also appear in this print version.

Printed by Curran Associates, Inc. (2018)

For additional information, please contact EDP Sciences – Web of Conferences at the address below.

EDP Sciences – Web of Conferences
17, Avenue du Hoggar
Parc d'Activité de Courtabœuf
BP 112
F-91944 Les Ulis Cedex A
France

Phone: +33 (0) 1 69 18 75 75
Fax: +33 (0) 1 69 28 84 91

contact-edps@webofconferences.org

Additional copies of this publication are available from:

Curran Associates, Inc.
57 Morehouse Lane
Red Hook, NY 12571 USA
Phone: 845-758-0400
Fax: 845-758-2633
Email: curran@proceedings.com
Web: www.proceedings.com

TABLE OF CONTENTS

Chairman of the Workshop – **Alexander Litvak,** Institute of Applied Physics, Russia

Organized by
Institute of Applied Physics, Russian Academy of Sciences http://iapras.ru/

GYCOM Ltd.
http://www.gycom.ru/

Sponsored by
Russian Foundation for Basic Research (grant 18-02-20056)
http://www.rfbr.ru/rffi/ru/

Program Committee

Litvak A.G.		IAP RAS, Nizhny Novgorod, Russia
Zapevalov V.E	Scientific Secretary, Deputy Chairman	IAP RAS, Nizhny Novgorod, Russia
Gusakov E.Z.		Ioffe Institute, St. Petersburg
Denisov G.G.		IAP RAS, Nizhny Novgorod, Russia
Roi I.N.		NRC "Kurchatov Institute", Moscow, Russia
Jelonnek J.		KIT, Karlsruhe, Germany
Thumm M.		KIT, Karlsruhe, Germany
Zohm H.	Co-Chairman	IPP, Garching, Germany
Wagner D.		IPP, Garching, Germany
Plaum B.		IGVP, Stuttgart, Germany

Organizing committee
(Institute of Applied Physics, Russia)

Glyavin M.Yu.
Sedov A.S.
Abramov I.S.
Fokin A.P.
Krapivnitskaya T.O.
Mocheneva O.S.
Morozova I.P.
Pavelyev A.B.
Orlov V.B.
Osharin I.V.

Scientific Secretary of the Meeting – **Vladimir Zapevalov,** Institute of Applied Physics, Russia

Statement of Peer review

In submitting conference proceedings to *Web of Conferences*, the editors of the proceedings certify to the Publisher that

1. They adhere to its **Policy on Publishing Integrity** in order to safeguard good scientific practice in publishing.
2. All articles have been subjected to peer review administered by the proceedings editors.
3. Reviews have been conducted by expert referees, who have been requested to provide unbiased and constructive comments aimed, whenever possible, at improving the work.
4. Proceedings editors have taken all reasonable steps to ensure the quality of the materials they publish and their decision to accept or reject a paper for publication has been based only on the merits of the work and the relevance to the journal.

Title, date and place of the conference

30th Joint Russian-German Meeting on ECRH and Gyrotrons, June 17-24, 2018
Nizhny Novgorod, Russia

Proceedings editor(s):

Dr. V.E. Zapevalov

Date and editor's signature

August 17, 2018

https://doi.org/10.1051/epjconf/201818701001

Modelling of EUV light sources based on microwave discharge in inhomogeneous flow of nonequilibrium plasma with multiply charged tin and xenon ions

I. S. Abramov, E. D. Gospodchikov, A. G. Shalashov

Institute of Applied Physics of Russian Academy of Sciences, Nizhny Novgorod, Russia, abramov@appl.sci-nnov.ru

Introduction

Extreme ultraviolet (EUV) lithography is key technology for further decrease of integral circuit (IC) scales [1]. Transition to smaller wavelength of exposing radiation requires generation of plasma with multiply charged ions of heavy elements emitting EUV light. Now IC manufacturers use sources based on laser evaporation of tin droplets that can provide 200–250 W of EUV light in 13.5 nm ±1% spectral band [2–4]. However experts believe that for significant improvement of IC form factor about 1 kW power is required that can hardly be obtained in laser produced plasmas [4].

A competitive alternative is microwave discharge supported by submillimeter wave radiation of modern high power gyrotrons. Duration of gyrotron pulses is much longer (50 μs to CW) and provided power is much higher (up to 1 MW) than those for lasers in use. These allow to support strongly radiating discharge in stationary mode. Moreover, microwave heating leads to nonequilibrium plasma with high-energy electrons and as a consequence high rates of electron impact ionization and excitation. Two concepts of such EUV light sources were developed in IAP: based on tin plasma flow propagating through the open magnetic trap and freely expanding flow of xenon plasma [5, 6]. Both concepts are tested in proof of principle experiments, but detailed theoretical study is required to explore physical constrains and prospects for these concepts. This study is in focus of present communication.

Theory

We use fluid equations taking into account multiple ionization by electron impact with the following assumptions:

- the discharge is stationary, thus we use stationary equations;
- the plasma is strongly inhomogeneous along the flow direction x, thus we assume plasma flow with the characteristics averaged over the flow cross section $S(x)$;
- difference in ion and electron masses allows to neglect the electron momentum flow;
- the temperature of electrons is much higher than ion one $T_e \gg T_i$ due to resonant heating, thus ion pressure is neglected;
- high electron temperature provides high electron thermal conductivity, thus T_e is conserved and enters the model as a parameter.

Under the assumptions above, fluid equations may be expressed as ordinary differential equations of ion fractions particle balance and the equation of total momentum balance [7, 8]

$$\tfrac{d}{dx}(n_j S u) = n_e S (k_{j-1} n_{j-1} - k_j n_j), \qquad (1)$$

$$\tfrac{d}{dx}\left(SMnu^2\right) + S\tfrac{d}{dx}(n_e T_e) = 0,\ n = \sum_{j=1}^{Z_{\max}} n_j, \qquad (2)$$

where n_j are densities of ions with charge-state $Z_j = j$ (i.e. with charge je), u is the flow velocity, n_e is the electron density followed from the quasi-neutrality of the plasma, k_j is j-th ionization constant, M is ion mass. As it follows from (1), the total flow $F = Snu$ is conserved and may be treated as another model parameter. Thus to close the problem it is necessary to set ion composition and flow velocity at some point inside the discharge. Boundary conditions for (1) and (2) are set taking into account presence of ion acoustic transition in the flow expanding into vacuum [8]. Formal mathematical difficulties with defining the transition location are treated as it is proposed in [9].

From the energy conservation, a power load that is required to support the high electron temperature may be estimated as sum over major channels of power losses from the discharge

$$P = AF\frac{Mu^2}{2}\bigg|_{x\to L} + \int (p_{\text{ion}} + p_{\text{rad}}) S dx. \qquad (3)$$

The first term represents the convective losses; here A is the coefficient depending on details of the plasma absorber located at L [8]. The second term describes the volumetric losses for ionization and line radiation of ions characterized with densities

$$p_{\text{ion}} - \sum_j E_j k_j n_e n_j, \quad p_{\text{rad}} = \sum_j \sum_l \sum_h F_{jhl} k^*_{jlh} n_e n_j$$

Here p_{ion} is the power spent to ionize a unit volume of multiply charged plasma, E_j is the ionization energy of j-th ion, $k_j n_e$ is the effective rate of electron impact ionization, p_{rad} is the power spent on line radiation and inner sums are taken over all possible terms of ground (l) and excited (h) electron configurations of j-th ion, E_{jhl} are the transition energies. Values k^*_{jlh} are effective excitation constants. Plasmas with densities close to 10^{16} cm^{-3} that may be obtained in studied discharges are not optically thin even for energetic EUV light quanta. So, for these conditions the constants k^*_{jlh} should be modified compared to optically thin limit by taking into account the effect of radiation trapping due to reabsorption of the emitted photons in the discharge volume. The method is developed by the authors and discussed in details in [10].

The theory was used to study prospects of both concepts of EUV light sources proposed in IAP. Here is the brief overview of the results.

Tin plasma in open magnetic trap

The first concept of the source based on additional heating of the vacuum arc plasma flow propagating in open magnetic configuration shown possibilities to generate large amounts of EUV power up to tens of kilowatts at present level of technological development (with modern gyrotrons and vacuum arc sources). However the main problem of these sources is that the emitting region is pretty long and it is nearly impossible to collect and focus all generated EUV light. Even modified magnetic configuration proposed by authors in [9] that significantly decreases the emitting area hardly can provide required sizes of about 1 mm. Fig. 1 illustrates modeling of the EUV light source that provides about 40 kW of the required radiation with about 20 % of microwave to EUV conversion efficiency. Emitting region size is about 1 cm.

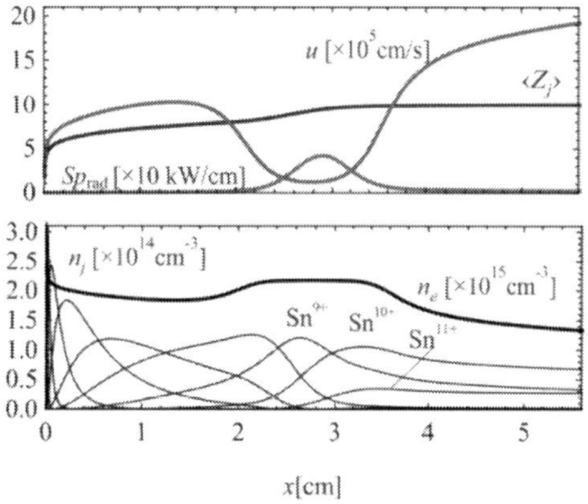

Fig. 1. Simulation of EUV-radiating discharge in tin plasma flow propagating through the open magnetic trap: variation along the trap axis of the flow velocity u (red line, top panel), average ion charge $\langle Z_j \rangle$ (blue line), linear power density Sp_{rad} of radiation in 13.5±1% nm EUV band (violet line), densities of ion fractions n_j (thin black lines, bottom panel), and electron density n_e (thick black line). Model parameters are: trap length $L = 5.6$ cm, mirror ratio $R = 9$, initial plasma diameter $d = 0.2$ cm, arc current $I_a = 2.9$ kA, electron temperature $T_e = 50$ eV, initial beam composition $n_1(0)/n(0) \approx 0.325$, $n_2(0)/n(0) \approx 0.567$, $n_3(0)/n(0) \approx 0.108$, and other components are negligible.

Discharge in freely expanding xenon jet

The second concept involves microwave discharge in freely expanding jet of heavy noble gases (presumably xenon). Such jet provides perfect (less than 1 mm) localization of the discharge but causes decrease in conversion efficiency due to increasing percentage of convective losses. To provide EUV light power of kW level by such a small discharge dense plasmas are required. However it is shown that gas puff rate and heating may be adjusted so that radiation trapping becomes a positive effect as more energetic radiation quanta are trapped at higher densities. Fig. 2 illustrates modeling of the EUV light source that provides about 5 kW of the required radiation with about 5 % of microwave to EUV conversion efficiency. Emitting region size is smaller than 1 mm.

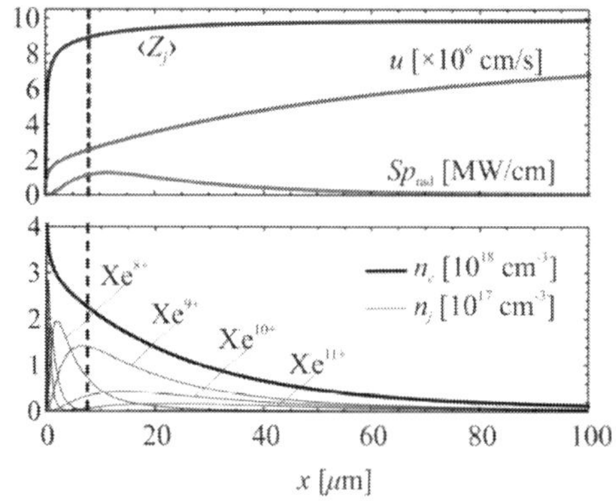

Fig. 2. Simulation of EUV-radiating discharge in freely expanding xenon jet: flow velocity u (red line, top panel, x is the distance from the nozzle), average ion charge $\langle Z_j \rangle$ (blue line), linear power density Sp_{rad} of radiation in 11.2±1% nm EUV band (violet line), densities of ion fractions n_j (thin black lines, bottom panel), and electron density n_e (thick black line). Parameters: nozzle diameter $d = 70$ μm, total flow $F = 2\times10^{19}$ s^{-1}, electron temperature $T_e = 100$ eV. Xenon is singly ionized at the nozzle.

Acknowledgements

The work was supported by Russian Foundation for Basic Research (Grants No. 17-02-00173 and No. 18-32-00419).

References

1. *Wagner, C., Harned, N.* Lithography gets extreme // Nature Photonics 2010. V. 4, P. 24–26.
2. *Mizoguchi, H., Nakarai, H., Abe, T. et al.* Performance of 250W high-power HVM LPP-EUV source // EUV Lithography VIII 2017, V. 10143. P. 101431J.
3. *Fomenkov, I., Brandt, D., Ershov, A. et al.* Light sources for high-volume manufacturing EUV lithography: technology, performance, and power scaling // Advanced Optical Technologies 2017. V. 6, P. 173 – 186.
4. *Moore, S.* EUV lithography finally ready for fabs // *IEEE Spectrum* 2018, V. 55. P. 46–48.
5. *Chkhalo, N. I., Golubev, S. V. et al.* Source for extreme ultraviolet lithography based on plasma sustained by millimeter-wave gyrotron radiation // J. Micro Nanolithogr. MEMS and MOEMS 2012, V. 11, P. 021123.
6. *Glyavin, M. Y., Golubev, S. V., Izotov, I. V. et al.* A point-like source of extreme ultraviolet radiation based on a discharge in a non-uniform gas flow, sustained by powerful gyrotron radiation of terahertz frequency band // Applied Physics Letters 2014, V.105(17), P. 174101.
7. *Abramov, I. S., Gospodchikov, E. D., Shalashov, A. G.* Formation of a multi-charged plasma in the directed gas flow // Radiophysics and Quantum Electronics 2016, V. 58. P. 914 – 933.
8. *Shalashov, A. G., Abramov, I. S., Golubev, S. V., Gospodchikov, E. D.* Theory of a stationary microwave discharge with multiply charged ions in an expanding gas jet // JETP 2016. V. 123, P. 219 – 230.
9. *Abramov, I. S., Gospodchikov, E. D., Shalashov, A. G.* Prospects of extreme ultraviolet radiation sources based on microwave discharge for high-resolution lithography // Physics of Plasmas 2017, V. 24, P. 073511.
10. *Abramov, I. S., Gospodchikov, E. D., Shalashov, A.G.* Potentials of EUV light source based on microwave discharge in expanding jet of dense xenon plasma // arXiv:1712.10026 [physics.plasm-ph] 2018.

https://doi.org/10.1051/epjconf/201818701002

Two-beam gyrotron with frequency multiplication

I. V Bandurkin[1], M. Yu. Glyavin[1], T. Idehara[2], and A. V. Savilov[1]

[1]Institute of Applied Physics, Russian Academy of Sciences, Nizhny Novgorod, Russia, iluy@appl.sci-nnov.ru
[2]Center for Development of Far-Infrared Region, University of Fukui, Fukui, Japan

Introduction

A well-known method of weakening the requirements for the magnitude of the operating magnetic field in gyrotrons is generation at high cyclotron harmonics. Gyrotrons have long demonstrated the ability to work at high cyclotron harmonics at frequencies above 1 THz. Nevertheless, the majority of gyrotrons using the conventional magnetron-injector gun (MIG) operate at a cyclotron harmonic not higher than the second. First of all, this is due to the aggravation of the selectivity problem, since a hollow electron beam produced in such a gun is coupled to many modes of the system. Parasitic oscillations at low cyclotron harmonics, as a rule, have higher coupling coefficients and therefore get an advantage in excitation. Partly, the problem of selectivity can be solved by the use of an axis-encircling electron beam, since the spectrum of the modes interacting with it is relatively rare. Nevertheless, the electron gun capable of formation of such a beam is rather complicated in comparison with the MIG, and the obtained electron current is limited due to the small area of the beam cross section. An alternative method of increasing the selectivity can be the use of the frequency multiplication effect, when the gyrotron is excited at the fundamental cyclotron resonance and simultaneously emits at one of the higher harmonics of the fundamental frequency. We propose a variant of such a gyromultiplier scheme, which can provide a relatively high generation efficiency in the terahertz range due to the use of two electron beams.

Two-beam gyromultiplier scheme

The main problem in the realization of a single-beam self-excited gyromultiplier is the fast saturation along the system of high-frequency currents at all cyclotron harmonics (Fig. 1a) [1-4]. In a two-cavity scheme, this leads to the operation of a second cavity at an overbunched beam with a low efficiency. A single-cavity scheme requires the use of special methods for providing a double resonance (at low and high cyclotron harmonics) in the same electrodynamic volume, which greatly complicates the system.

We suggest a two-cavity scheme in which the electron overbunching problem is solved by using two electron beams (Fig. 1b). One of these beams is strongly coupled with the mode of the first cavity at the fundamental frequency and ensures its self-excitation. The second beam has a relatively weak coupling with this low-frequency mode due to the correct choice of its radius. Therefore, its modulation in the first cavity is also weak, and the maximum bunching in it is achieved only in the second cavity, where, at this time, it is strongly coupled with the mode at a high harmonic (Fig 1c).

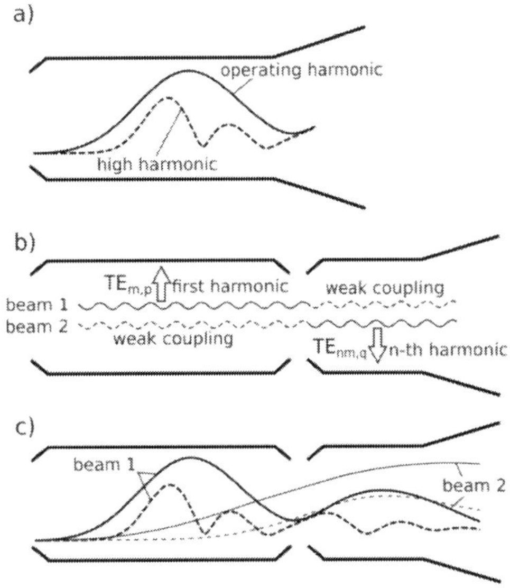

Fig. 1. Typical distributions of high-frequency current at various harmonics of the operating frequency in a conventional gyrotron (a), proposed two-cavity two-beam gyromultiplier scheme (b) and distribution of high-frequency currents in it (c).

Selection of operating modes and beams radii

For an example design we have chosen operation at first and third cyclotron harmonics, at the frequencies close to 400 GHz and 1.2 THz, correspondingly. A characteristic feature of gyromultipliers is that the azimuthal indices of operating modes should be multiples of the same coefficient as their frequencies and cyclotron harmonic numbers. On the one hand, this additionally increases the selectivity of the harmonic generation, but, on the other hand, this requirement complicates the selection of operating modes. It is simplest to choose the system parameters in the case of axisymmetric modes with a zero azimuthal index. For instance, when a $TE_{0,4}$ and $TE_{0,13}$ modes are selected at the first and third cyclotron harmonics, the beam radii can be chosen so that each of the beams is well coupled with its mode, and simultaneously is near the zero of excitation of the other mode (Fig. 2a). Moreover, since the starting current at the fundamental harmonic is low, it is reasonable to put the larger part of the total electron beam current to the outer beam destined for the high harmonic operation. In the discussed design, the ratio of electron currents in inner and outer beams equals to 1:9. Nevertheless, when choosing the second beam radius equal to 0.44 mm in order to get larger coupling coefficient at the third cyclotron harmonic, the system becomes vulnerable to parasitic excitation of the $TE_{-2,4}$ mode at the fundamental cyclotron resonance. Indeed, while the co-rotating mode $TE_{2,4}$ has exactly the same dependence of the coupling factor on the beam radius as the $TE_{0,4}$ mode, the counter-rotating mode $TE_{-2,4}$ has maximum of the coupling coefficient at the position of the high-current outer beam. Since the cutoff frequency of the $TE_{2,4}$ mode is 0.988 of the one of the $TE_{0,4}$ mode, the desired operation can be easily suppressed by the $TE_{-2,4}$ mode excited in the regime of BWO (Fig. 2b). Thus, it is sensible to choose a larger outer beam radius of 0.82 mm, so as to weaken the coupling factor of the parasitic mode (Fig. 2c).

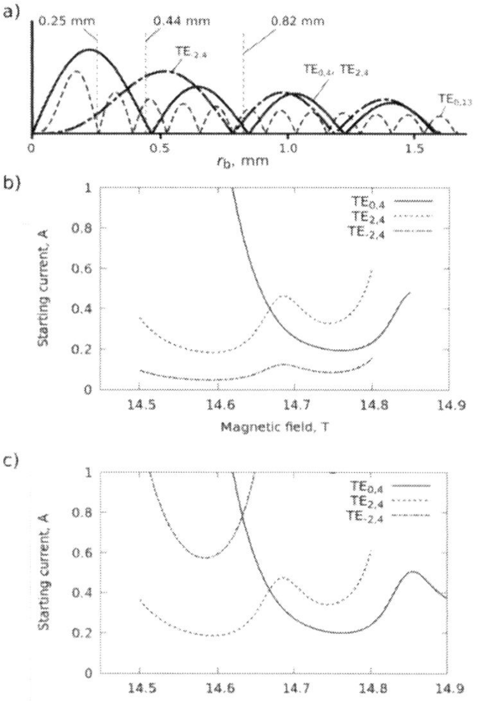

Fig. 2. Dependence of the coupling factors of operating and spurious modes on the beam radius as well as transverse positions of the inner (0.25 mm) and outer (0.44 mm or 0.82 mm) beams (a), and starting currents of desired and parasite excitations for the outer beam radii of 0.44 mm (b) and 0.82 mm (c).

Numerical simulations

In the simulations, both the beams were supposed to have the same electron energy of 22.8 keV, the pitch factor of 1.3, and the 20% spread in transverse electron velocities. The total operating current was equal to 1 A, 0.1 A being put in the inner electron beam and 0.9 A in the outer one. For the radii of first and second cavity sections equal to 1.6 mm and 1.665 mm correspondingly, the maximum output power at the third cyclotron harmonic at the frequency of 1.2 THz amounted to approximately 6 W (Fig. 3). Fig. 4 shows the distribution of operating fields and rf-currents along the cavity in this case.

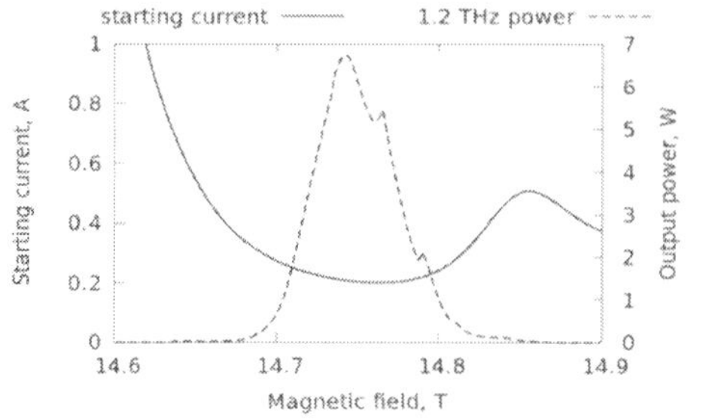

Fig. 3. Dependence of the third cyclotron harmonic output power on the magnetic field for the 1 A total beam current (simulations).

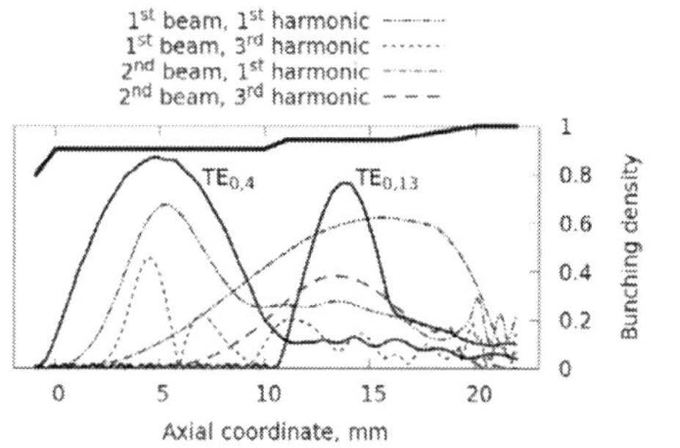

Fig. 4. Distributions of the high-frequency currents along the sectioned cavity for the case of maximum output power.

Conclusion

A new scheme of a two-resonator frequency gyromultiplier based on the use of two coaxial hollow electron beams is proposed and theoretically investigated. Efficient generation in such a scheme is achieved due to a special choice of the beam radii, when one of the beams provides excitation of the first cavity at the fundamental cyclotron resonance, while the other excites the second cavity at a high cyclotron harmonic. For such a gyromultiplier, the calculations predict the possibility of terahertz generation at the third cyclotron harmonic with a watt-level power.

The work is supported by the Russian Science Foundation, Project No. 14–12–00887.

References

1. *Bandurkin, I.V., Bratman, V.L., Savilov, A.V.* Frequency multiplication in gyrotron autooscillators // Tech. Phys. Lett. 2006. V. 32, No. 1. P. 84–87.
2. *Savilov, A.V., Nusinovich, G.S.* On the theory of frequency-quadrupling gyroklystrons // *Phys. Plasm.* 2007. V. 14, No. 5, P. 053113.
3. *I.V. Bandurkin, V.L. Bratman, G.G. Denisov, Yu.K. Kalynov, A.V. Savilov, A.W. Cross, W. He, K. Ronald, A.D.R. Phelps.* Single-cavity gyromultipliers // Terahertz science and technology. 2008. V. 1, No. 3, P. 169.
4. *I.V. Bandurkin, V.L. Bratman, A.V. Savilov, S.V. Samsonov, A.B. Volkov.* Experimental study of a fourth-harmonic gyromultiplier. Phys. Plasmas. 2009. V. 16, No. 7, P. 070701.

Concurrent operation of 10 gyrotrons at W7-X - experience and improvement opportunities

H. Braune, K.J. Brunner, H. P. Laqua, S. Marsen, D. Moseev, F. Noke, F. Purps, N. Schneider, T. Schulz, T. Stange, P. Uhren, F. Wilde and the W7-X team at IPP, IGVP Stuttgart and KIT Karlsruhe
Max-Planck-Institut für Plasmaphysik, Garching and Greifswald, Germany

Abstract – **At the stellarator Wendelstein 7-X (W7-X) Electron Cyclotron Resonance Heating (ECRH) is the main heating system for steady-state operation and was the only available heating system during the operation phase OP1.2a in 2017. The ECRH, equipped with 10 operational 140 GHz gyrotrons [1], was used for different tasks at W7-X such as wall conditioning, controlled plasma start-up from the neutral gas up to steady state plasma control and different heating scenarios such as X2-mode and O2-mode heating as well as current drive (ECCD). The operational experiences and improvement opportunities of the ECRH plant will be discussed.**

I. Introduction And Background

THE recent W7-X operation phase OP1.2a was performed with 10 140 GHz Gyrotrons in total. W7-X is currently equipped with an inertially cooled divertor with an input energy limit up to 80 MJ. The main goal of this operation phase was to explore and prepare operation scenarios for steady-state high performance operation in future campaigns, when W7-X is fully equipped with water-cooled plasma facing components. The W7-X operation was powered by ECRH exclusively. The available total port through microwave power was more than 7.2 MW as shown in Fig. 1. Stationary discharges up to 30 s were achieved only limited by the maximum divertor energy load. The long discharges were also used to demonstrate current control and bootstrap current compensation by ECCD.

The ECRH transmission line system was supplemented by two remote steering launchers, an important concept for the future fusion reactor [2] which was tested at W7-X for the first time for plasma heating and current drive.

II. Preparation of the ECRH Installation

The individual power characteristics (acceleration voltage vs. output power) of each gyrotron were measured. Based on these power characteristics for the expected pulse length up to 30s, the W7-X main control real time system has only one parameter to control the output power of each gyrotron. By these means, the power variation capability during discharges is achieved.

The previous main power supply set up has the capability to provide 10 gyrotrons, two of them in parallel arrangement. The couple of gyrotron that operated in parallel was changed with the objective of equivalent output power at the same cathode voltage feed. Two gyrotrons with comparable behaviour pattern delivered by Thales Electron Devices (TED) were selected. The cathode voltage feed of different type gyrotrons separately, one delivered by TED and the other one delivered by Communications & Power Industries (CPI), leads to a gain of 100 kW microwave output power at least per gyrotron compared to the first approach when these gyrotrons were operated in parallel based on their point of installation within the ECRH set up. The modification was performed by changing of high voltage connection cables against each other due to the move of gyrotrons would require much more effort.

Three of the existing main power modules were equipped with a new controller in connection with new software release by the manufacturer Amperon. The voltage and current thresholds (50A / 65kV) are replaced by a slightly higher power threshold (currents up to 70A with 50kV output voltage will be achievable -> sufficient for 1.5 MW gyrotron).

III. Heating Scenarios - Achieved Main Plasma Parameters

The W7-X plasmas were generated and sustained with the 2nd harmonic EC heating in the well-absorbed extraordinary polarization (X2) at densities below the X-mode cut off of 1.2×10^{20} m^{-3}. Current control and bootstrap current compensation by ECCD was demonstrated within the long discharges up to 30 s. The rotational transform iota and thus the confinement and stability were intensely changed by localised ECCD.

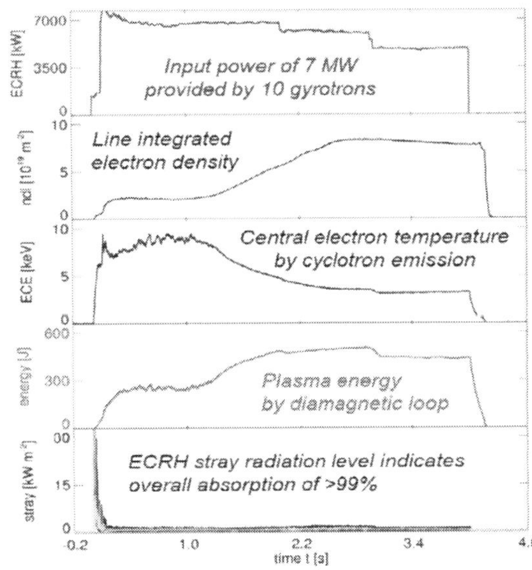

Fig. 1: ***Helium plasma discharge with 10 used gyrotrons, at around 2 s and 3 s the power stepped down by switching off of gyrotrons***

In order to avoid beam reflection at densities close to the cut off and above that are relevant for a future fusion reactor, a dedicated triple beam path heating scenario with O-mode polarisation has been developed. Target plasma was generated by at least three gyrotrons with the X-mode polarisation. Afterwards, the remaining gyrotrons were switched on in O-mode and the start-up gyrotrons were changed to O-mode as well. A density of 1.4×10^{20} m^{-3} was achieved in combination with pellet injection.

Due to the first path O2 mode absorption of only 70% a triple beam path was developed as shown in Fig. 2 based on individual holographic reflection tiles for each beam and a polished steel panel at the low field side. The O2 mode absorption was increased up to 90% with these three passes through the EC-resonance in the plasma centre.

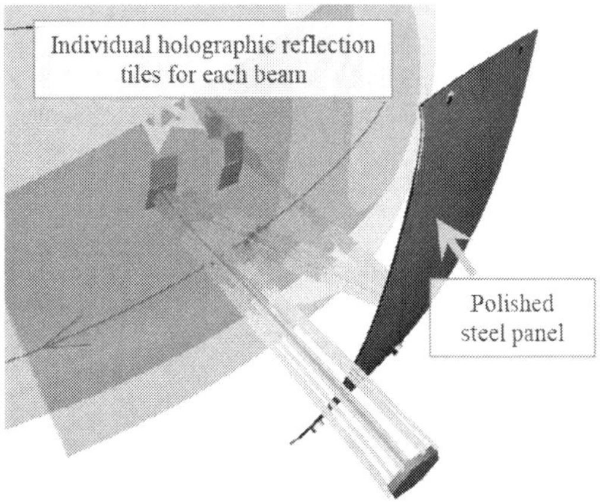

Fig. 2: Example of triple beam path for one (Alpha1) beam-line

IV. New gyrotron operation modes

The W7-X gyrotrons were operated not at the maximum output power level during the previous operation phases in order to avoid gyrotron interlocks caused by instabilities.
A new software solution based on the fast gyrotron control system tries to keep the gyrotron in operation by acceleration voltage reduction for 500µs. This solution [3] uses the capabilities of the fast control FPGA in connection with the achievable slew rates in the range of 0.4 kV/µs of the main and the body power supplies. The fast gyrotron control system monitors the microwave output power based on microwave sensitive diodes and if the output power drops below a certain threshold the gyrotron acceleration voltage would be reduced about 12 kV as shown in Fig 3 for both high voltages separately.

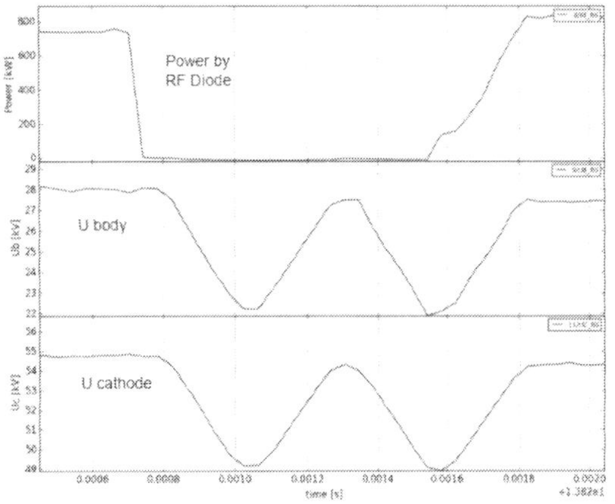

Fig. 3: RF power, body voltage, cathode voltage vs. time

Two fast acceleration voltage reduction cycles were required in this case to achieve approximately the same microwave output power as before by recovering the gyrotron working mode.
The gyrotron interlock system turns the gyrotron off if the output power does not come back within 2 ms.
This kind of mode recovery allows a more reliable gyrotron operation close to the maximum output power limit. The gain of average power would be likely 100 kW per gyrotron.

V. Outlook

The collector sweep system of the W7-X gyrotrons delivered by TED needs to improve in order to reduce the collector surface temperature slightly more as already achieved especially at the bottom region. This sweep system consists of two different coil arrangements, Vertical Field Sweep and Transverse Field Sweep, which are fed by single power supplies separately. A synchronised operation of both sweep power supplies will lead to a more smooth power distribution on the bottom of the collector surface and will avoid the shift of the electron strike line at the top of the collector toward to the cover plate that are not sufficiently cooled.

The W7-X ECRH installation offers the opportunity to enhance the amount of gyrotrons up to 12. The main power supply upgrading is currently under construction. First of all the power supply is going to provide 12 gyrotrons simultaneously. Furthermore, a slightly higher power threshold replaces the voltage and current thresholds of 50A and 65kV. Currents up to 70A with 50kV output voltage will be achievable and that are sufficient for prospective 1.5 MW gyrotrons which are currently under development at different manufacturers [4] [5].

References

[1] V. Erckmann et al. "ECRH and W7-X – an intriguing pair" Radio Frequency Power in Plasmas, Proc. 20 Topical Conference, Sorrento, Italy, AIP Conference Proceedings 1580 (2013) page 542
[2] W. Kasparek et al. "Remote-Steering Launchers for the ECRH system on the Stellarator W7-X", EPJ Web of Conferences 87, 04005 (2015)
[3] F. Wilde et al. "Measurements of Satellite Mode Activity and Automated Mode Recovery in 140 GHz Wendelstein 7-X Gyrotrons", 20th Joint Workshop on Electron Cyclotron Emission (ECE) and Electron Cyclotron Resonance Heating (ECRH), May 14-17, 2018 Greifswald, Germany, Book of abstracts page 24
[4] G. Denisov " New Developments of Megawatt Power Gyrotrons in Russia", 20th Joint Workshop on Electron Cyclotron Emission (ECE) and Electron Cyclotron Resonance Heating (ECRH), May 14-17, 2018 Greifswald, Germany, Book of abstracts page 18
[5] J. Jelonek et al. "Towards Advanced Fusion Gyrotrons: 2018 Update on Activities within EUROfusion", 20th Joint Workshop on Electron Cyclotron Emission (ECE) and Electron Cyclotron Resonance Heating (ECRH), May 14-17, 2018 Greifswald, Germany, Book of abstracts page 19

Acknowledgement:
This work has been carried out within the framework of the EUROfusion Consortium and has received funding from the Euratom research and training programme 2014-2018 under grant agreement No 633053. The views and opinions expressed herein do not necessarily reflect those of the European Commission.

EPJ Web of Conferences 187, 01004 (2018) https://doi.org/10.1051/epjconf/201818701004
30th Joint Russian-German Meeting on ECRH and Gyrotrons

Low loss $MgAl_2O_4$ ceramics for terahertz windows

S.V. Egorov[1], A.A. Sorokin[1], I.E. Ilyakov[1], B.V. Shishkin[1], V.V. Parshin[1], S.S.Balabanov[2], A.V.Belyaev[2]

[1]Institute of Applied Physics RAS, Nizhny Novgorod, Russia
[2] G.G. Devyatykh Institute of Chemistry of High-Purity Substances RAS, Nizhny Novgorod, Russia

This paper describes fabrication of polycrystalline magnesium aluminate spinel by microwave sintering and hot pressing and investigation of its dielectric properties in the terahertz frequency range. The significant interest in this material is associated with its high performance properties including the impact strength and the microhardness, the possibility to operate at high temperatures and to withstand exposure to radiation. In addition, using an appropriate fabrication method the spinel can be made optically transparent. Along with that it has been shown that $MgAl_2O_4$ ceramics obtained by microwave sintering of high-purity ultrafine powders has a sufficiently high transparency in the millimeter-wave range (the dielectric loss tangent $\tan \delta < 6.0 \cdot 10^{-4}$ at a frequency of 200 GHz) [1]. Such ceramics can be used both in microelectronics, for example, for the manufacture of microwave ceramic resonators, and in high-power electronics – to create output windows in high-power vacuum devices of the millimeter-wave range.

In general, the processing route of low-loss ceramics fabrication follows the common practice of ceramic technology, viz. powder production, compaction and sintering. A special attention is paid to the purity and dispersity of the synthesized powder materials. In our case the initial powders were prepared by the sol-gel method. This method of synthesis and deep purification makes it possible to obtain high-purity, nanosized powders. We used two approaches for the ceramics consolidation in this work: microwave sintering and hot pressing of the powder compacts. The absence of heating elements in the microwave furnace ensures pure vacuum conditions, which makes it possible to obtain ceramics with low dielectric losses. It should be noted that one of the problems of spinel sintering is a considerable loss of magnesium (Mg) during long-term exposure to high temperature in vacuum. The capability of achieving high heating rates inherent in microwave processing allows retaining the stoichiometry of the spinel. As for hot pressing, this method provides additional densification of the sintered material and results in achieving high transparency in the visible and near-infrared range. However, the hot pressing method of spinel consolidation requires introduction of sintering additives, such as lithium fluoride (LiF), which can affect the dielectric characteristics of the final material.

The microwave sintering of the compacts was carried out in a gyrotron-based system operating in CW mode at a frequency of 24 GHz with a variable output power up to 6 kW. Hot pressing of the compacts was carried out using a uniaxial press at a temperature of 1600 °C and a pressure of 35 MPa.

To study the dielectric properties, high-density ceramic samples were produced with a diameter of 25 mm and a thickness of 2 mm. The samples fabricated by microwave sintering were either pure spinel or 1 % Y_2O_3-doped. The samples produced by uniaxial hot pressing with the LiF sintering additives were optically transparent. The dielectric characteristics were studied in the frequency range 0.05÷0.3 THz using a spectrometer based on an open high-quality factor Fabry –Perot resonator, and in the range 0.6 ÷3.3 THz using the method of terahertz time-domain spectroscopy. The results have shown that the refractive index of the samples varies weakly from 2.85 to 2.92 in the studied frequency range (Fig. 1). The refractive index of optically transparent ceramics is slightly higher than that of opaque samples (fabricated by microwave sintering). This fact may be associated with a higher density of optically transparent materials.

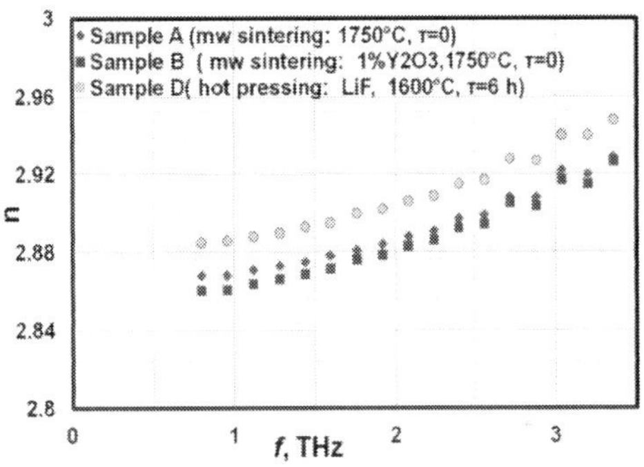

Fig. 1. The frequency dependencies of the refractive index of $MgAl_2O_4$ ceramics

The frequency dependencies of the dielectric losses of the ceramics are presented in Fig. 2. It can be seen that the value of tan δ increases significantly from $1.5 \cdot 10^{-4}$ to 10^{-2} within the frequency range from 0.05 to 3.3 THz. The pure material has the lowest value of the dielectric losses. The presence of sintering additives in the ceramics slightly increases the electromagnetic absorption in the material. A comparative analysis has shown that the dielectric losses of the spinel in the terahertz frequency range are comparable in magnitude to the losses of sapphire [2] and pure polycrystalline aluminum oxide (Fig. 2) [3], while they are lower than those of the boron nitride ceramics [4].

In conclusion, the results of this study have demonstrated that pure $MgAl_2O_4$ ceramics obtained by microwave sintering exhibits relatively low dielectric losses in the terahertz frequency range and can be used, in particular, for the fabrication of output windows of terahertz vacuum electronic devices. The ceramics obtained by hot pressing is also transparent in the visible and near-infrared range and can be used for the fabrication of multispectral protective windows.

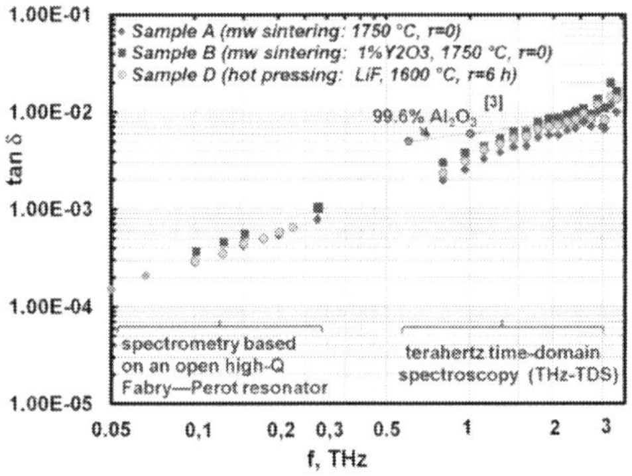

Fig. 2. The frequency dependencies of the dielectric loss tangent of $MgAl_2O_4$ ceramics

Acknowledgment

The study of the microwave sintering of MgAl2O4 ceramics was supported by Russian Foundation for Basic Research, grant No. 16-08-00736. The study of the terahertz dielectric properties was supported by the Russian Foundation for Basic Research and the Government of the Nizhny Novgorod region, grant No. 18-42-520015.

References

1. *Egorov, S.V., Bykov, Yu.V., Eremeev, A. G., et.al.* Millimeter-wavelength radiation used to sinter radiotransparent $MgAl_2O_4$ ceramics // Radiophysics and Quantum Electronics 2017, V. 59, No. 8–9, P. 690-697.
2. *Grischkowsky, D., Keiding, S.,van Exter, M. and Fattinger, Ch.* Far-infrared time-domain spectroscopy with terahertz beams of dielectrics and semiconductors // J. of Opt. Soc. Am. B 1990, V. 7, No. 10, P. 2006–2015.
3. *Rajab, K. Z., Naftaly, M., Linfield, H., et.al.* Broadband dielectric characterization of aluminum oxide (Al_2O_3) // J. Micro. and Elect. Pack. 2008. V 5, P. 101–106.
4. *Naftaly, M., Greenslade, P. J., Miles, R. E., Evans, D.* Low loss nitride ceramics for terahertz windows // Optical Materials 2009. V. 31, P. 1575–1577

Augmenting microwave irradiation in MAS DNP NMR samples at 263 GHz

Armin Purea[1], Benjamin Ell[2], Christian Reiter[1], Fabien Aussenac[3], Frank Engelke[1],

[1]Bruker Biospin GmbH, Rheinstetten, Germany, frank.engelke@bruker.com
[2]Karlsruhe Institute of Technology, Karlsruhe, Germany, [3]Bruker Biospin France, Wissembourg, France

Abstract

The magnetic microwave field strength and its detailed spatial distribution in magic-angle spinning (MAS) nuclear magnetic resonance (NMR) probes capable of dynamic nuclear polarization (DNP) is investigated by numerical simulations with the objective to augment the magnetic microwave amplitude by structuring the sample in the mm and sub-mm range and by improving the coupling of the incident microwave beam to the sample. As it will be shown experimentally, both measures lead to an increase of the microwave efficiency in DNP MAS NMR.

As in almost all magnetic resonance experiments, in particular in electron paramagnetic resonance (EPR) and in NMR experiments, the magnitude of the rf magnetic field is a crucial parameter. In DNP MAS NMR electron spin polarization is transferred to nuclear spins with the effect of enhancing nuclear spin polarization by one or two orders of magnitude above the Boltzmann polarization in solid samples at temperatures around 100 K. In order to achieve the electron spin – nuclear spin polarization transfer, unpaired electron spins in radical molecules embedded in an NMR spectroscopic sample of interest are irradiated by an EPR resonant microwave beam (frequency, *e.g.*, 263 GHz, linearly polarized, gyrotron source). For the specific DNP effect to take place, the amplitude, phase, and polarization distribution of the microwave field during the scattering process in the sample material and influenced by the surrounding hardware components must be known and, if possible, optimized to achieve high DNP enhancement factors. We have studied two strategies to increase the DNP transfer efficiency.

The first strategy relies on embedding dielectric particles (*e.g.*, potassium bromide, KBr, grains) into the sample [1,2], the former with a dielectric constant higher than the sample material (tetrachloroethane, TCE, plus low concentration biradicals, TEKPOL). The powder grains have size distributions between 200 and 600 μm, are randomly distributed in the TCE frozen solution with volume filling factors of 0% (no particles), 22% and 63%. The sample and the particles are contained in a cylindrical 3.2mm or 1.3mm outer diameter MAS rotor (allowing fast spinning of the sample with rates of tens of kilohertz) made of zirconia ceramics or sapphire. The MAS rotor is surrounded by a solenoidal rf coil being part of an NMR resonant circuit, *e.g.*, for ^{1}H NMR at 400 MHz for an external static field of 9.4 T. At that external field strength, the electron spins resonate at 263 GHz. The incident linearly polarized microwave beam (HE_{11} mode, waist *c.a.* 3 mm) leaves a corrugated waveguide (inner diameter 7.6 mm), propagates through the coil, the MAS rotor wall and the sample. On that path of propagation, the beam is multiply diffracted or scattered by the coil, the rotor wall and the sample. The numerical microwave field simulations show that already the presence of the rotor wall (zirconia or sapphire) causes a complicated diffraction pattern of the beam inside the sample with a distribution of left and right-handed elliptical polarization and in the absence of dielectric particles a spatially periodic distribution of magnetic microwave amplitudes. The presence of dielectric particles maintains the general occurrence of elliptical polarization in the sample but moreover yields a random distribution of field amplitudes originating from the multiple scattering of the beam on the randomly sized and distributed particles. Furthermore, in the case of dielectric particles the spatially averaged (over the sample volume) field amplitude B_1 increases by a factor of 1.5 as compared to the case of absent particles, see Fig. 1. We may attribute the improved experimental DNP efficiency by a factor 2 to 3 being proportional to B_1^2 to this increase of B_1. The detailed impact of elliptical polarization is currently still under investigation.

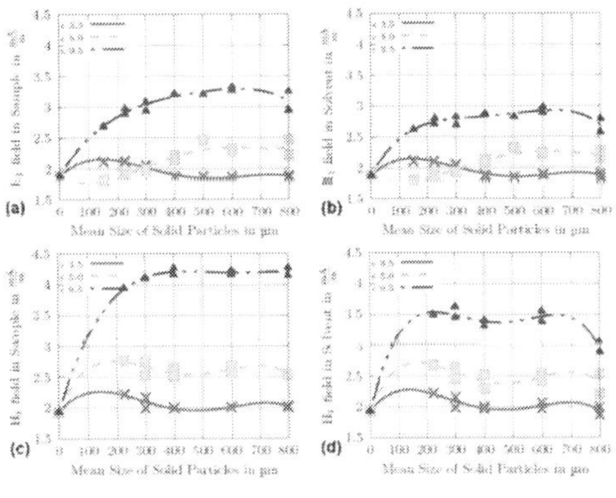

Fig. 1. Magnetic microwave field H_1 *vs.* average particle size for dielectric particle volume filling factor 22% (a,b) and 63% (c,d) averaged over the total sample volume (particles and TCE) (a,c) and TCE alone (b,d) for various dielectric constants ε = 3.5, 5.0, and 9.5. ε = 3.5 is equal to the dielectric constant of TCE, hence particles with such an ε are "dielectrically indistinguishable" from TCE. ε = 5 corresponds to KBr, ε = 9.5 to sapphire.

The second strategy [3] aims at an improved coupling of the incident microwave beam to the sample by uti-

https://doi.org/10.1051/epjconf/201818701005

lizing a focusing lens at the aperture of the 7.6 mm corrugated waveguide and an additional short waveguide of only 2 mm inner diameter guiding the beam very close to the sample, Fig. 2.

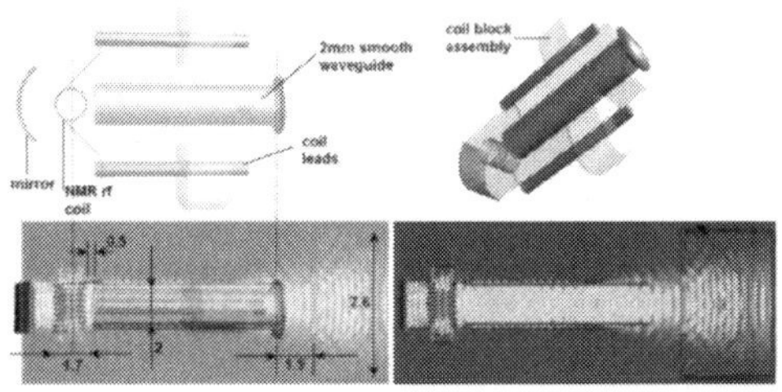

Fig. 2. Improved microwave beam coupling to the rf coil, MAS rotor, and DNP sample for a 1.3mm MAS DNP probe at 263 GHz. The beam enters from the right side after passing a focusing lens (diameter 7.6 mm), enters the small waveguide (inner diameter 2 mm), and propagates through an aperture of that waveguide close to the rf coil, MAS rotor and sample. Color coding: red – high microwave amplitude, yellow – intermediate amplitude, blue – low amplitude.

Tab. 1. Magnetic microwave amplitudes without and with waveguide coupler compared with the amplitude achievable in a fundamental mode EPR cavity (Q = 500)

System	$B_1/P^{1/2}$ ($\mu T/W^{1/2}$)	$\gamma_e B_1/P^{1/2}/2\pi$ ($MHz/W^{1/2}$)
MAS 1.3 mm DNP probe without waveguide coupler	19	0.5
MAS 1.3 mm DNP probe with waveguide coupler	66	1.8
EPR E780 spectrometer with fundamental mode resonator	1272	35.8

As summarized in Tab. 1 the insertion of the focusing lens in the aperture of the large corrugated waveguide and the insertion of the short, small-diameter waveguide increases the magnetic microwave amplitude B_1 in the sample by a factor of 3. For comparison, the last line in Tab. 1 indicates the attainable magnetic field amplitude in an EPR cavity (quality factor 500, 20 mW input power, 90° pulse 50 ns) at 263 GHz, which is much higher because of the optimum resonating structure. Such a more ideal structure is hard to achieve for MAS DNP probes for the reasons of the various hardware boundary conditions for MAS DNP as compared to pure EPR.

^{1}H DNP 1.3 mm MAS NMR experiments at 263 GHz with ca. 7 W incident beam power with water/glycerol/proline/ AMUPol yielded DNP enhancements of about 150 and 350 for the setup without and with waveguide coupler, a significant improvement of the DNP efficiency. Both strategies, the embedding of dielectric particles into the sample matrix and the waveguide coupler can be combined and produce ^{1}H DNP enhancements up to 500 in TCE/KBr/TEKPOL, a value not so far anymore from the theoretically attainable value of 650 for ^{1}H DNP NMR.

References

1. *Kubicki, D.J., Rossini A.J., Purea, A., Zagdoun, A., Ouari, O., Tordo P., Engelke, F., Lesage, A., Emsley, L.*, Amplifying dynamic nuclear polarization of frozen solutions by incorporating dielectric particles, // J. Am. Chem. Soc. 2014. V. 136. P. 15711-15718.

2. *Kubicki, D.J., Casano, G., Schwarzwälder M., Abel, S., Sauvee C., K. Ganesan, K., Yulikov, M., Rossini, A.J., Coperet, C., Lesage, A., Purea A., Engelke, F., Tordo, P., Ouari, O., Emsley, L.*, Rational design of nitroxide biradicals for efficient cross-effect dynamic nuclear polarization", in: 57th ENC, Pittsburgh, 2016.

3. *Purea, A., Reiter, C., Dimitriades, A.*, NMR-MAS-Probenkopf mit optimiertem MAS-DNP-Spulenklotz für schnelle Probenrotation, Patent submitted to Deutsches Patent und Markenamt, DE102018204913, 29 March 2018.

 https://doi.org/10.1051/epjconf/201818701006

Design and Experimental Test Of 250 GHz/300 kW/CW Gyrotron

G. Denisov[1,2], A. Fokin[1], M. Glyavin[1], A. Kuftin[1], M. Morozkin[1], V. Malygin[1], M. Proyavin[1], A. Sedov[1], E.Soluyanova[1,2], E. Sokolov[1,2], A. Tsvetkov[1], E. Tai[1,2], V. Zapevalov[1]

[1]Institute of Applied Physics of the Russian Academy of Sciences, Nizhny Novgorod, Russia, ap.fokin@mail.ru
[2]GYCOM Ltd., Nizhny Novgorod, Russia

One of the most important applications of modern gyrotrons is the plasma heating and control of the current in controlled thermonuclear fusion facilities. 170-GHz 1MW-power continuous regime gyrotrons with efficiency about 50% is developed for the ITER project in several countries, including Russia. Now ITER is under intensive construction. The concept of DEMO fusion power plant includes the upgrade of the heating and current drive systems. The main points are the increase of heating frequency up to 230 GHz and higher and increase in the total microwave power up to 50 MW [1]. In order to satisfy new conditions it is important to solve many scientific and engineering problems, and high-frequency high-power gyrotrons are now undergoing development in many groups [2, 3]. Here we present the design and experimental results of a 250 GHz continuous-wave (CW) gyrotron developed jointly by IAP RAS and GYCOM Ltd.

Gyrotron design

The main limiting factor for the design of the gyrotron is the hot bore diameter of available cryomagnet. For cryogen-free magnet JMTD-10T100 (JASTEC Inc., Japan) the 100 mm hot bore limited the size of an electron-optical system and quasi-optical converter. As a result, the $TE_{19.8}$ mode was chosen as the operating one with a cavity radius of 9.34 mm. The cavity length was determined by maximum thermal loading on the cavity wall, which for ideal copper cannot exceed 1.5 kW/cm^2. Based on the numerical simulation of the electron-wave interaction efficiency, the cavity length was set as 10 mm with nominal beam current for CW regime 12.5 A. The electron beam is formed by a diode-type magnetron injection gun, and recuperation of residual electron energy is performed by depress collector. The total accelerating voltage U_0 = 55 kV, with negative cathode potential, grounded collector and body voltage of 20 kV. The possibility of stable single-mode oscillation of the operating mode was demonstrated by numerical modeling in self-consistent model with efficiency up to 40%.

The tube is equipped with Denisov-type quasi-optical converter, which provides the transformation of the operating mode to the Gaussian beam and separation of wave and electron beams. It consists of shaped waveguide, parabola, four guiding mirrors and synthesized mirror for direction of the wave beam in the output window. For initial production and short pulse tests of the tube the simple boron nitride (BN) window with diameter of 66 mm was used. Its thickness was optimized to minimize the reflections at an operating frequency 250 GHz. The future CW tests of the gyrotron assume the replacement of the BN window by a CVD diamond window.

Operating frequency	250 GHz
Operating mode	$TE_{19.8}$
Accelerating voltage, U_0	55 kV
Body voltage	20 kV
Beam current (CW)	12.5 A
Cavity radius	9.34 mm
Cavity length	10 mm
Beam radius	3.93 mm

Fig. 1. The general view of the 250 GHz gyrotron installed in the cryomagnet

Experimental test

The experimental investigation of the produces tube were performed in a pulse-periodic regime, with pulse duration of 40 μs and repetition rate of 10 Hz. The first tests were performed without electron energy recuperation, using pulsed power supply developed and manufactured in IAP RAS. In experiments, the used accelerating voltage was up to 60 kV with beam currents up to 20 A. Also, an additional coil was installed in the cathode region of the tube for fine tuning of the magnetic field on the cathode.

The measurement system included NorthStar PVM-5 high voltage probe for measurement of the accelerating voltage, Rogowski coil with sensitivity of 50 mV/A for measurement of beam current and a wa-

 https://doi.org/10.1051/epjconf/201818701006

ter calorimetric dummy load for power measurements. The output power was calculated based on measured study-state average power in pulse-periodical regime with taking into account duty cycle. The frequency measurements were performed using resonant-cavity wave meter.

The maximum output power measured in experiment reached 330 kW at the operating frequency 249.74 GHz with total efficiency more than 30% without energy recuperation. The measured dependence of output power and efficiency on the current of the electron beam s presented on the fig.2. For nominal beam parameters of 55 kV and 15.5 A the output power was about 200 kW, which is well corresponding with theoretical estimations on the design stage.

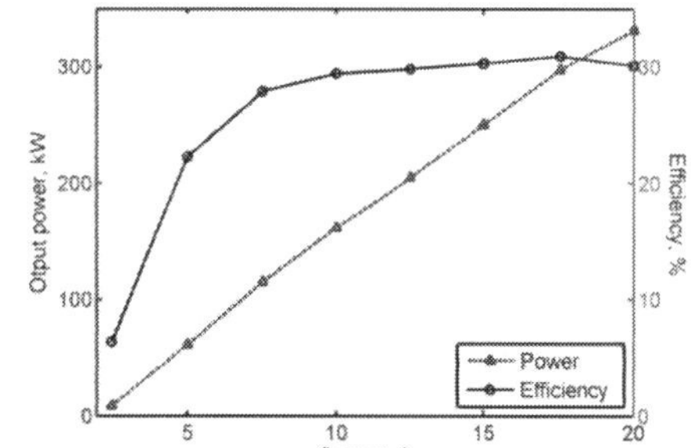

Fig. 2. Dependence of output power and efficiency on beam current

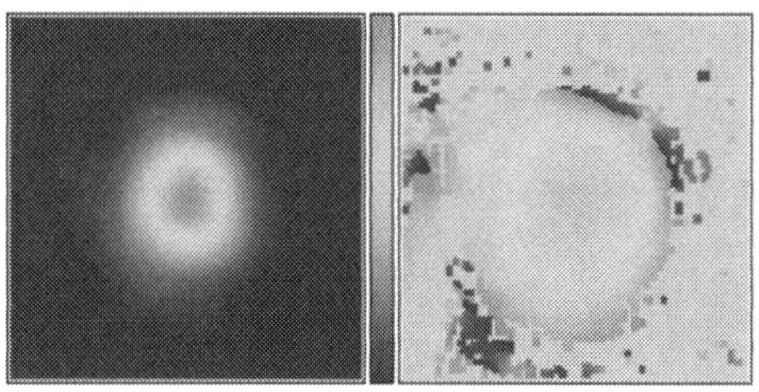

Fig. 3. Measured amplitude and reconstructed phase of the beam

The measurements of microwave beam distribution were performed using the infrared camera technique [4] and presented in fig.3. Resulting content of the Gaussian beam exceeds 98%.

The gyrotron was successfully used for initiation of localized gas discharge. The main result of such experiments is detection of plasma emission in the EUV spectral ranges of 13-17 nm (for xenon) and 18-50 nm (for xenon and argon) [5].

Conclusion

The 250 GHz gyrotron has been developed, manufactured and experimentally tested. The microwave power up to 330 kW with 30% efficiency without collector depression was obtained in 40 μs pulses.

This work was supported by Russian Scientific Fund (RSF), project № 14-12-00887.

References

1. *G.Granucci et al.* The EC-system of EU DEMO: concepts for a reactor heating system // EPJ Web of Conferences 149, 03003 (2017).
2. *R.Ikeda et al.* Development of multi-frequency gyrotron for ITER and DEMO at QST // 41st International Conference on Infrared, Millimeter, and Terahertz Waves (IRMMW-THz 2016) Copenhagen, Denmark.
3. *J.Jelonnek et al.* Design considerations for future DEMO gyrotrons: A review on related gyrotron activities within EUROfusion // Fusion Engineering and Design, 123, 241-246 (2017).
4. *S.O.Kuznetsov, V.I.Malygin.* Determination of gyrotron wavebeam parameters // Int. J. of Infrared and Millimeter Waves, 12, 11, 1241-1252 (1991).
5. *A.G.Shalashov et al* Observation of extreme-ultraviolet light emission from an expanding plasma jet multiply-charged argon or xenon ions // Appl. Phys. Lett. (submitted).

EPJ Web of Conferences **187**, 01007 (2018)
30th Joint Russian-German Meeting on ECRH and Gyrotrons
https://doi.org/10.1051/epjconf/201818701007

Data Transmission Using Gyrotron Radiation as a Carrier

A. Fokin, A. Sedov, A. Tsvetkov

[1]Institute of Applied Physics of the Russian Academy of Sciences, Nizhny Novgorod, Russia, ap.fokin@mail.ru

The idea of advancement of data transmission systems to the sub-THz region is actively discussed in the scientific and engineering community [1,2]. The available frequencies are limited by the absorption in the atmosphere, so the range of 200-300 GHz seems very promising. But for this frequency range the complication in the implementation of communication systems is the lack of medium-power (up to 1 kW) sources with wide tuning capabilities. In this paper we present the first experiments using the sub-THz gyrotron as the radiation source for data transmission.

Gyrotron and control system

The experiments were carried out on the 263 GHz gyrotron for spectroscopy and media diagnostics with output power up to 1 kW [3]. The control of output power and frequency was performed by variation of anode voltage of the triode-type magnetron-injection gun. Such system was already used for frequency stabilization of the gyrotron [4] and is capable of modulating parameters of gyrotron radiation with modulation frequency up to 1 MHz, defined by the time constant of the circuit. The anode voltage control unit is capable of variation of anode voltage directly proportional to the input signal with amplitude of 0 – 10 V.

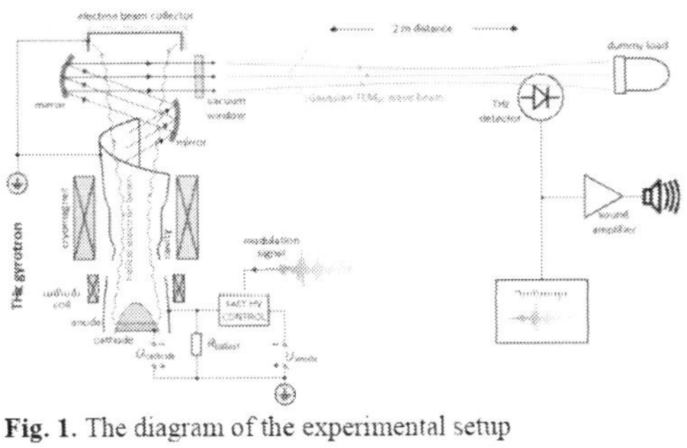

Fig. 1. The diagram of the experimental setup

Experimental test

The diagram of the experimental setup is presented in the fig. 1. In the first experiments the simple amplitude modulation of the gyrotron radiation was used. Detection of the modulated sub-THz radiation at the distance of 2 m was performed by an ultrafast THz detector produced by TeraSense company. We used two types of input data signals: an analog sound signal, taken from the loudspeaker exit of the car stereo, and the pseudo-random bit sequence, formed by the arbitrary waveform generator Agilent 33521B.

In the experiment we demonstrated the possibility of exact transmission of audio signal, using the gyrotron radiation as a carrier, which is justified by the analysis of the resulting oscilloscope traces of the input and resulting signal and by direct playback of the received signal. For the digital data transmission, we demonstrated the maximum rate of about 1.5 Mbit/s. The simple estimates of the system show the possibility of the increase of transmission rates up to 1 Gbit/s with minor upgrade of the control unit.

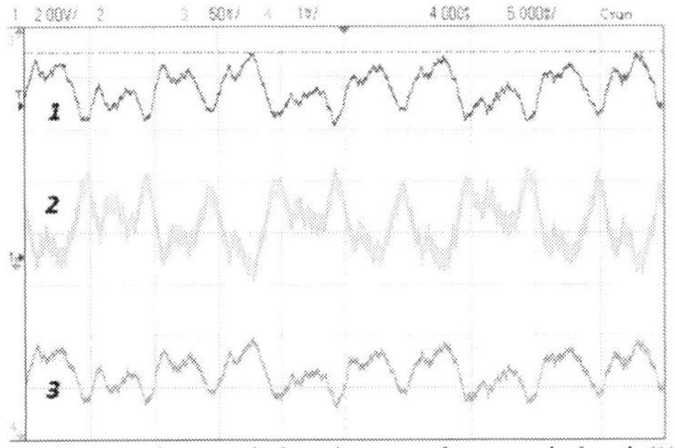

Fig. 2. Sound transmission: input analog sound signal (1) signal from detector (2) and anode voltage (3)

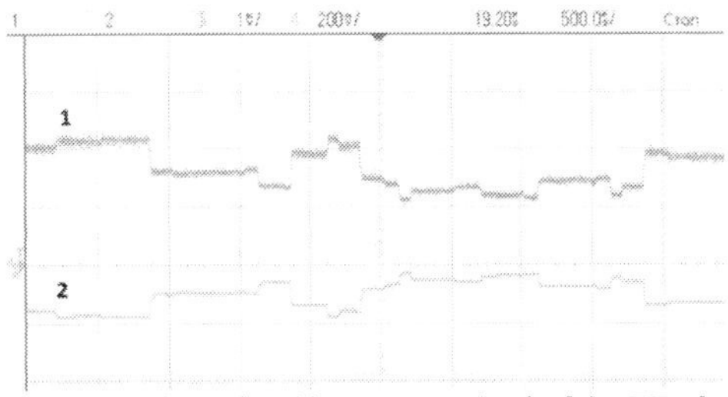

Fig. 3. Pseudo-random bit sequence: signal of the THz detector (1) and input sequence (2)

Conclusion

The possibility of data transmission using the modulation of the gyrotron radiation was demonstrated. The precise transmission of audio signal and 1.5 Mbit/s data rate for digital signal were obtained.

The work was partially supported by the Russian Federation President Grant No. MK-3452.2017.8

References

1. *T. Schneider et al.,* Link budget analysis for Terahertz fixed wireless links // IEEE Trans. on Terahertz Science and Technology, 2, 250 (2012)
2. *T. Kürner and S. Priebe,* Towards THz communications-status in research, standardization and regulation, J. Infrared, Millimeter, and Terahertz Waves, 35, 53-62 (2014)
3. *M. Glyavin et al.* Experimental tests of 263 GHz gyrotron for spectroscopy applications and diagnostic of various media // Review of Scientific Instruments, 86, No. 5, 054705 (2015)
4. *A. Fokin et al.* High-power sub-terahertz source with a record frequency stability at up to 1 Hz // Scientific Reports, 8, 4317 (2018)

Development of Terahertz-Range Planar Gyrotrons with Transverse Energy Extraction Operating at Cyclotron Harmonics

N. S. Ginzburg[1], T. Idehara[2], Yu.V. Zaslavsky[1], I. V. Zotova[1], M. Yu. Glyavin[1], A. S. Sergeev[1], A.M. Malkin[1], I. V. Zheleznov[1]

[1]Institute of Applied Physics of the Russian Academy of Sciences, Nizhny Novgorod, Russia, ginzburg@appl.sci-nnov.ru
[2]Research Center for Development of Far-Infrared Region, University of Fukui, Japan

In the last few years, a considerable progress has been achieved in the development of terahertz-range gyrotrons [1-4]. At the frequency of 1 THz, the radiation power amounts several kilowatts in conventional gyrotrons with tubular helical beams formed by magnetron injection guns [1-3] and up to hundreds watts in large-orbit gyrotrons (LOG) with axis-encircling electron beams [4]. All these experiments were performed in the pulse regime. Recently, terahertz-range radiation at the second cyclotron harmonic was achieved. In the latter experiment, a superconductive magnet was used. The typical feature of such magnets is fairly large diameter of the warm bore of 5-10 cm. At the same time, the transverse cross-sections of conventional terahertz gyrotrons with cylindrical resonator are quite limited by several millimeters (Fig.1a). It is caused by problems with mode selection that restricted admissible waveguide radius and correspondingly the driving current. In order to provide starting conditions, one should increase interaction length above an optimal value that together with rather small cross-section results in substantial Ohmic losses. According to simulations for experimental conditions corresponding to [5], the Ohmic losses amount 80% of radiation power. Obviously, the improvement of mode selection is the key issue in the further development of short-wavelength gyrotrons.

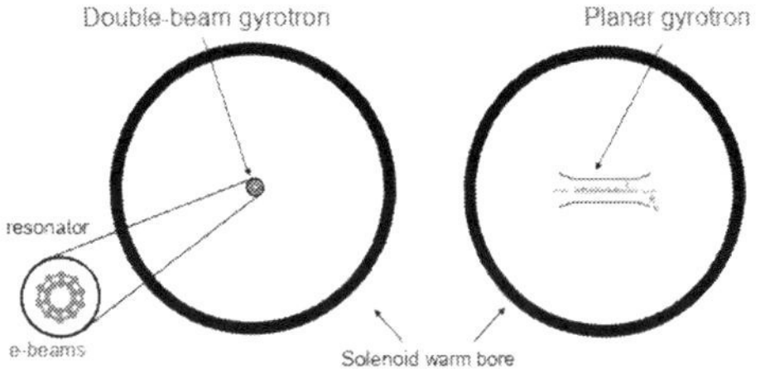

Fig. 1. Transverse cross-section of (a) a double-beam and (b) a planar gyrotrons in the scale of the warm bore of the JMTD15T52 cryomagnet.

For drastic increase in the output power of short-wavelength gyrotrons, we suggest to use the planar scheme with a sheet electron beam and transverse (with respect to the electrons translation velocity) electromagnetic energy extraction (Fig.1b). The main advantage of this scheme comparing to the conventional cylindrical geometry is the possibility of effective mode selection over the open transverse coordinate in combination with radiation out-coupling, which leads to substantial reduction of Ohmic losses [6]. It is important to note that in the existing cryomagnets, the warm bore is sufficient for a drastic increase of cross-sections of terahertz-range gyrotrons and provides enough space for installation of additional reflectors required for arrangement of transmission of generated radiation in the direction of collector.

An additional advantage of the planar scheme is the peculiarity associated with excitation of odd ($s = 1,3...$) and even ($s = 2,4...$) cyclotron harmonics. Under assumption that the sheet electron beam is injected in the middle of the cavity between the plates, the interaction at odd cyclotron harmonics occurs only for resonator modes with odd transverse indexes $n = 1,3...$, while interaction at even harmonics occurs only for modes with even transverse indexes $n = 2,4...$. Moreover, for example, for operation at the second harmonic it is beneficial to use an even resonator mode with indexes n equal to doubled even number. In this case interaction at second harmonic will be not accompanied by simultaneous excitation of a lower order mode at the first cyclotron harmonic due to the coupling factor for the 1st harmonic with an even mode is equal to zero. From the other hand, for excitation at the odd cyclotron harmonic the number of the resonator mode should not be dividable by s. For example, for $s = 3$, the resonator mode number may be $n = 5,7,11...$. In this case the parasitic mode at the 2nd cyclotron harmonic is not excited.

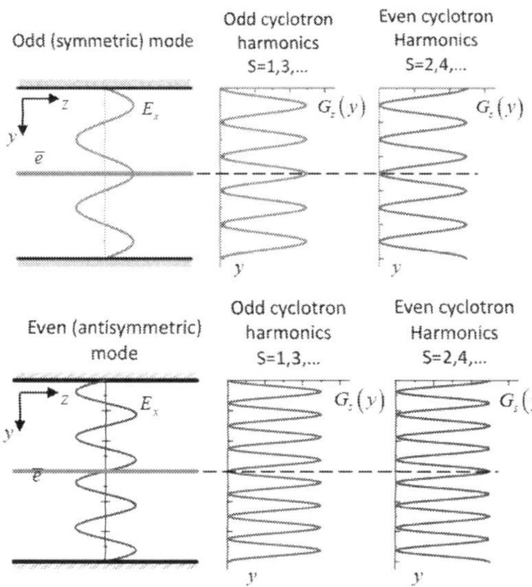

Fig. 2. The coupling factors $G(y)$ in a planar gyrotron.

Further results of simulations are presented for a 750 GHz second harmonic planar gyrotron with following parameters: accelerating voltage of 30 kV, electron current of 2 A, pitch-factor of 1.2, distance between plates of 4 mm, plates width and length were 8 and 17 mm, correspondingly. Simulations were carried out in the frame of averaged multiple modes averaged approaches:

$$s\frac{\partial a_n}{\partial \tau} + i\frac{\partial^2 a_n}{\partial X^2} + i\frac{\partial^2 a_n}{\partial Z^2} + (i\Delta_n + \sigma)a_n = \frac{I_n}{2\pi}\int_0^{2\pi} p^s d\vartheta_0,$$

$$\frac{\partial p}{\partial \tau} + \frac{g^2}{4}\frac{\partial p}{\partial Z} + ip\left(|p|^2 - 1\right) = -\sum_n a_n \left(p^*\right)^{s-1}.$$

Here a_n is the amplitude of the $TE_{n,1}$ mode, p is the transverse electron momentum, Δ_n is the cyclotron resonance detuning, I_n is the parameter proportional to electron current and the coupling factor $G_n(y)$, σ is the parameter of Ohmic losses proportional to the skin depth.

In suggested scheme of gyrotron the diffractional extraction of radiation is realized in transverse x direction. Correspondingly at the edges of interaction space the nonreflecting boundary conditions may be applied. In the direction of translational motion of electrons the waveguide is closed by cut-off necks we can apply zero boundary condition along z coordinate.

In simulations competitions of operating $TE_{20,1}$ at the 2nd cyclotron harmonic and parasitic $TE_{9,1}$ modes at the 1st cyclotron harmonic was taken into account (Fig.3). The mode $TE_{20,1}$ doesn't excite due to corresponding factor $G_{10}(y) = 0$.

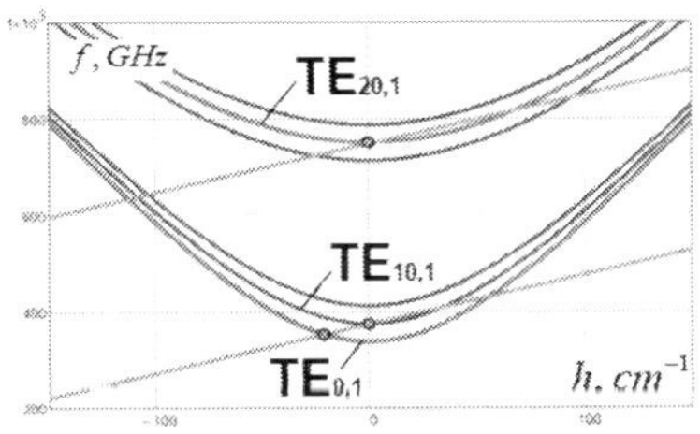

Fig. 3. Dispersion diagram of competing modes.

Results of simulations are presented in Fig.4. Figure 4a demonstrates the competition scenario. One can see that the excitation of the parasitic mode $TE_{9,1}$ is suppressed. As a result the single mode generation with excitation of the $TE_{20,1}$ mode is realized. According to simulations the transverse efficiency is of $\eta_\perp$ ~12% (Fig. 4b) and the total efficiency is η ~7%. Taking into account that for chosen parameters about 30% of total radiation power is dissipated in Ohmic losses the output power is estimated of about 3 kW. It should be noted that in comparison with conventional gyrotrons, the suggested planar scheme demonstrates the high selectivity alongside with significantly lower Ohmic losses.

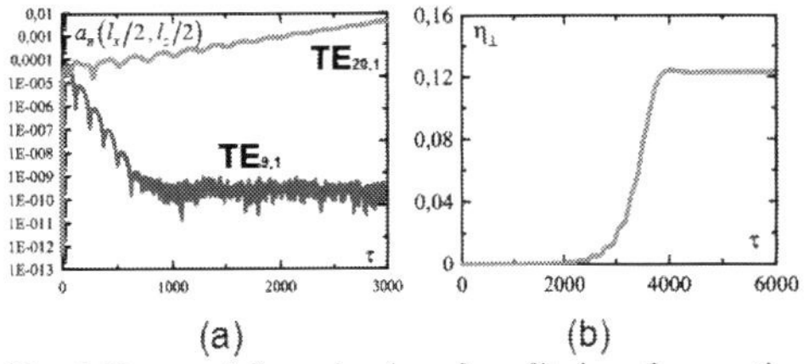

Fig. 4. Temporal dependencies of amplitudes of competing modes (a) and the transverse electron efficiency (b).

The work was supported by the Russian Science Foundation under grant No. 18-19-00704.

References

1. *Idehara T., Tsuchiya H., Watanabe O., Mitsudo, S.* The first experiment of a THz gyrotron with a pulse magnet // Int. J. Infrared and Millimeter Waves 2006. V. 27. No. 3. P. 319-331.

2. *Glyavin M.Yu., Luchinin A.G., Golubiatnikov G.Yu.*, Generation of 1.5-kW, 1-THz coherent radiation from a gyrotron with a pulsed magnetic field // Phys. Rev. Lett. 2008. V. 100. No. 1 Art.no. 015101.

3. *Notake T., Saito T., Tatematsu Y., Fujii A., Ogasawara S., Agusu La, Ogawa I., Idehara T., Manuilov V.N.*, Development of a novel high power sub-THz second harmonic gyrotron // Phys. Rev. Lett. 2009. V. 103. Art.no. 225002.

4. *Bratman V.L., Kalynov Yu.K., Manuilov V.N.*, Large-orbit gyrotron operation in the terahertz frequency range // Phys. Rev. Lett. 2009. V.102. Art.no. 245101.

5. *Idehara T., Glyavin M., Kuleshov A., Sabchevski S., Manuilov V., Zaslavsky V., Zotova I., Sedov A.*, A novel THz-band double-beam gyrotron for high-field DNP-NMR spectroscopy // Rev. of Sci. Instr. 2017 V.88. Art.no. 094708.

6. *Ginzburg N.S., Zotova I.V., Sergeev A.S., Zaslavsky V.Yu., Zheleznov I.V.*, High-power terahertz-range planar gyrotrons with transverse energy extraction// Phys. Rev. Lett. 2012. V. 108. Art.no. 105101.

2018 Status on KIT Gyrotron Activities

John Jelonnek[1,2], G. Aiello[3], K. Avramidis[1], G. Gantenbein[1], G. Grossetti[3], S. Illy[1], Z. C. Ioannidis[1], J. Jin[1], P. Kalaria[1], A. Marek[1], I. Gr. Pagonakis[1], T. Rzesnicki[1], S. Ruess[1,2], T. Ruess[1], T. Scherer[3], M. Schmid[1], D. Strauss[3], M. Thumm[1,2], F. Wilde[1,4], C. Wu[1], A. Zein[1]

[1]IHM, [2]IHE, [3]IAM, Karlsruhe Institute of Technology (KIT), 76131 Karlsruhe, Germany
[4]Max-Planck-Institut für Plasmaphysik, Teilinstitut Greifswald, 17491 Greifswald, Germany

Introduction

Karlsruhe Institute of Technology (KIT) is focusing on the development of gyrotron oscillators and related components for Electron Cyclotron Resonance Heating (ECRH) and Current Drive (CD) systems for magnetically confined nuclear fusion plasmas. It includes the participation in the developments for W7-X [1], ITER [2], TCV [3] and a future EU DEMOnstration power plant (DEMO) [4]. In 2017, KIT has been invited by IAP to collaborate in a joint DFG-RSF project on fundamental research for future generation of ultrashort pulses in the millimeter and submillimeter wave bands for spectroscopy and diagnostics in various areas based on passive mode-locking in gyrodevices with nonlinear electron cyclotron absorber in the feedback loop. This presentation is summarizing the 2018 status on the KIT gyrotron activities.

Status on the KIT fusion gyrotron developments

The fig. 1 shows the major fusion gyrotrons in development at KIT. That are the W7-X 140 GHz 1 MW CW series gyrotron, the ITER 170 GHz 1 MW short-pulse (ms) prototype, and, finally, the KIT 170 GHz 2 MW short-pulse (ms) pre-prototype.

In February 2016 the full operation of the initial ECRH system of the stellarator W7-X started successfully. It is using 9 series gyrotrons produced by Thales Electron Devices (TED), France and 1 gyrotron manufactured by CPI, USA. In 2018, all 10 gyrotrons were operated and close to 8 MW EC heating power was injected into the plasma already.

Since the successful start of W7-X, KIT has been focusing on the development of the first EU 170 GHz 1 MW hollow-cavity gyrotron for ITER. This development is coordinated by Fusion for Energy (F4E) and is ongoing within the European GYrotron Consortium (EGYC). The first industrial CW prototype bases on the 1 MW short-pulse (SP) modular pre-prototype gyrotron that has been manufactured and tested at KIT earlier [5]. The experiments with the CW industrial prototype gyrotron are running in two phases. In 2016, the first phase was completed at KIT. Major results were an RF output power of above 0.9 MW in short-pulses. The measured Gaussian mode content of the RF output beam was higher than 97 %. Pulses with a duration of 180 s (limited by the max. possible pulse length of the high-voltage power supply at KIT) delivered an RF output power higher than 0.8 MW with 38 % efficiency (in depressed collector operation) [6]. The second phase of the experiments is currently ongoing at EPFL-SPC, Lausanne. In parallel, at KIT, the target is the optimization of the RF output power and efficiency. Experiments with a modified SP modular pre-prototype are ongoing therefore. Various different schemes of depressed collector operation have been tested. The target is to achieve a total gyrotron efficiency higher than 50 %. Moreover, different designs of beam tunnels between electron gun and cavity are under consideration.

Fig. 1. The KIT fusion gyrotron family: KIT 170 GHz 2 MW short-pulse (ms) pre-prototype; W7-X 140 GHz 1 MW CW series gyrotron; ITER 170 GHz 1 MW short-pulse (ms) prototype (from left to right).

Based on the successful start of operation of the W7-X gyrotron, already in 2015, EPFL-SPC, Lausanne, Switzerland, invited KIT to collaborate in the physical design of a new TCV dual-frequency 84/126 GHz gyrotron with 1 MW RF output power and 2 s pulse length. The boundary condition in this development was to keep the technical design as close as possible to the W7-X gyrotron development. Nevertheless, several improvements, e.g. a triode-type magnetron injection electron gun, have been introduced. Just before the Russian-German Workshop 2018, EPFL operated the new TCV gyrotron for the first time. The RF output beam shows a promising Gaussian mode content. Additionally based on the original W7-X design, and, in parallel to the TCV development, KIT was looking into different options for a possible 140 GHz 1.5 MW gyrotron upgrade for W7-X. A possible physical design based on an upgrade to the TE28,10 cavity mode were presented.

In frame of the EUROfusion project, the basic research and development of a multi-megawatt gyrotron for future EU DEMO is ongoing. The 170 GHz, 2 MW short-pulse coaxial-cavity pre-prototype at KIT has already exhibited excellent performance in pulses of several ms [7]. The next step towards DEMO is to prove experimentally its capability for long-pulse operation. The coaxial-cavity pre-prototype was upgraded with new, water-cooled components and experiments targeting at 100 ms pulses are starting. To keep the development path towards the DEMO gyrotron as fast and cost-effective as possible, the design of a 2 MW, 170/204 GHz coaxial gyrotron has been initiated using the existing 170 GHz coaxial-cavity gyrotron as a starting point.

The target of significant better than 60 % efficiency for the DEMO gyrotron implies the development of advanced, Multi-Stage Depressed Collectors (MDCs). Given that in the gyrotron the electrons are facing a strong magnetic field at the collector region, the required separation of electrons according to their energy, necessary for MDC operation, is quite challenging. Investigations on different MDC concepts resulted in a configuration, which is based on the **ExB** drift concept [8] and adopts helical electrodes. The simulation results, carefully taking into account the influence of secondary electrons, demonstrate a gyrotron efficiency of higher than 63 % with a two-stage collector.

To allow gyrotron frequency step-tunability the CVD-diamond Brewster-angle window is under research and development. A major challenge is to produce a large diamond disk of 180 mm diameter and 2 mm thickness. New manufacturing technologies are under investigation to increase the diameter up to the required value. Growth tests at the site of the industrial partner (Diamond Materials GmbH, Freiburg, Germany) were initiated with promising first results.

Vital to the future gyrotron R&D is the new Fusion Long Pulse Gyrotron Lab (FULGOR), a test stand able to support the development of up to 4 MW continuous wave gyrotrons at frequencies up to 240 GHz [9]. Within the EUROfusion project, the basic specification for a super-conducting 10.5 T cryogen-free magnet for FULGOR has been prepared. The magnet is expected to be delivered by mid-2019.

Contributions to the Joint DFG-RSF Project - Generation of Ultra-Short Microwave Pulses -

In a joined DFG-RSF project of IAP and IHM, the generation of a periodic sequence of powerful short RF pulses is studied [10]. Simulations for the design of a simple quasi-optical feedback system have been performed at KIT using the KIT code KARLESSS [11]. Full-wave PIC simulations including the interaction between electrons and electromagnetic waves are performed with the advanced simulation program PICLas [12], developed by the Institute of Aerodynamics and Gas Dynamics (IAG) at the University of Stuttgart. The required high-order hexahedral meshes are generated with the mesh-generator HOPR, also developed by the IAG.

Acknowledgement

This work is partly supported by Fusion for Energy under the Grants F4E-GRT-432 and F4E-GRT-553 to the European Gyrotron Consortium (EGYC), F4E-OPE-458 to KIT and F4E-OPE-447 to TED. EGYC is a collaboration among SPC, Switzerland; KIT, Germany; HELLAS, Greece; IFP-CNR, Italy. Other parts of this work are partly carried out within the framework of the EUROfusion Consortium and are receiving funding from the Euratom research and training programme 2014-2018 under grant agreement No 633053. The views and opinions expressed herein do not necessarily reflect those of F4E and the European Commission. A part of this work was carried out using the HELIOS supercomputer system at Computational Simulation Centre of International Fusion Energy Research Centre (IFERC-CSC), Aomori, Japan, under the Broader Approach collaboration between Euratom and Japan, implemented by Fusion for Energy and QST. Part of the simulations was performed on the EUROfusion High Performance Computer (Marconi-Fusion). The joint IAP-IHM research on the generation of ultrashort pulses is ongoing in frame of the joint RSF-DFG project (Je 711/1-1).

References

1. *Erckmann V. et al.* ECRH and W7-X, an Intriguing Pair // AIP Proc. 20th Top. Conf. on RF-Power in Plasmas, Sorrento, Italy, P2.06 (2013).
2. *Alberti S. et al.* Dual-frequency, 126/84 GHz, 1 MW gyrotron for the upgrade of the TCV EC-system // Proc. IRMMW-THz Conf., Hong Kong (2015).
3. *Darbos C. et al.* Status of the ITER EC System // Proc. IEEE PPPS, San Francisco, USA, P5-22 (2013).
4. *Tran M. Q. et al.* EU DEMO Heating and Current Drive: Physics and Technology // Proc. 26th Fusion Energy Conf., Kyoto, FIP/7-7 (2016).
5. *Rzesnicki T. et al.* Recent experimental results of the European 1 MW, 170 GHz short-pulse gyrotron prototype for ITER // Proc. IRMMW-THz Conf., Hong Kong (2015).
6. *Ioannidis Z. et al.* First CW Experiments with the EU ITER 1 MW, 170 GHz Industrial Prototype Gyrotron // Proc. IEEE IVEC, London (2017).
7. *Rzesnicki T. et al.* 2 MW, 170 GHz Coaxial-Cavity Short-Pulse Gyrotron — Single Stage Depressed Collector Operation // Proc. IRMMW-THz Conf., pp. 1-2 (2014).
8. *Pagonakis I. et al.* A New concept for the collection of an electron beam configured by an externally applied axial magnetic field // IEEE Trans. Plasma Sci., vol. 36, no. 2, pp. 469-480 (2008).
9. *Schmid M. et al.* The 10 MW EPSM modulator and other key components for the KIT gyrotron test facility FULGOR // Fus. Eng. Des. (2017).
10. *Ginzburg, N. S., et al.* Generation of a periodic sequence of powerful ultrashort pulses in a traveling wave tube with bleachable absorber in the feedback loop." Technical Physics Letters, pp. 836-839 (2015).
11. Marek A. et al., Simulation of electromagnetic fields scattered from arbitrary shaped electric conductors // EPJ Web of Conferences 149, 04016 (2017).
12. Munz, C.-D. et al. Coupled particle-in-cell and direct simulation Monte Carlo method for simulating reactive plasma flows //Comptes Rendus Mécanique (2014).

Impact of plasma turbulence on linear mode conversion of quasi-optical wave beams in toroidal magnetic traps

T. A. Khusainov[1], A. G. Shalashov[1], E. D. Gospodchikov[1], and A. Köhn[2]

[1]Institute of Applied Physics of Russian Academy of Sciences, Nizhny Novgorod, Russia, hta@appl.sci-nnov.ru
[2]Universität Stuttgart, IGVP, Stuttgart, Germany

Introduction

Linear transformation of the ordinary (O) and extraordinary (X) electromagnetic modes in the vicinity of the plasma cutoffs in magnetized plasma is a basis for advanced schemes of microwave heating and diagnostic of over-dense plasmas (where the central density exceeds the cutoff) in magnetic fusion research [1]. On the other hand, plasma density fluctuations are typical of magnetic fusion experiment.

In this work, we investigate the influence of density fluctuations on the process of linear O-X mode coupling in non-one-dimensional magnetized plasma, using analytical approaches and full-wave numerical simulations.

We limit our studies to a two-dimensional model of the coupling region, having in mind that taking into account the actual three-dimensional structure of the magnetic field and the curvature of the magnetic surfaces in tokamaks and stellarators does not lead to a qualitative change in the equations describing the linear transformation [3].

Analytical model

We assume that the primary influence of density fluctuations on the O-X mode coupling efficiency is due to a small-angle scattering along the whole trace of a microwave beam, while the influence of the fluctuations on the tunneling process itself (inside the coupling region) can be neglected [2]. Such scattering leads to an additional phase modulation of the wave beam, but not to the refraction or reflection of the beam as a whole. Thus, we arrive at the following statement of the problem – a *random wave beam* that has experienced the influence of fluctuations along the path falls on the *regular domain* of the O-X mode coupling.

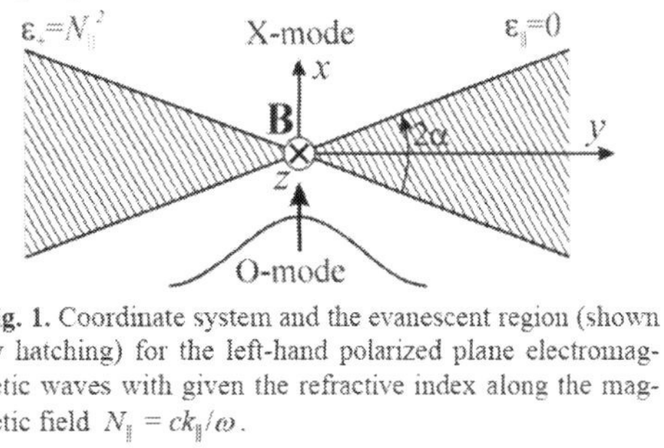

Fig. 1. Coordinate system and the evanescent region (shown by hatching) for the left-hand polarized plane electromagnetic waves with given the refractive index along the magnetic field $N_\parallel = ck_\parallel/\omega$.

The structure of the coupling region for the two-dimensional model is shown in Fig. 1. O-wave tunnels through the evanescent area formed by or intersecting the cutoff planes $\varepsilon_+(x,y) = N_\parallel^2$ and $\varepsilon_\parallel(x,y) = 0$. Here ε_+ and $\varepsilon_\parallel$ are the components of the dielectric permittivity tensor in the Stix representation,

$$\varepsilon_\parallel = 1 - n_e/n_{cr}, \quad \varepsilon_+ = 1 - (n_e/n_{cr})/(1 + B/B_{res}),$$

where $n_{cr} = m_e\omega^2/(4\pi e^2)$ is the critical plasma density for the wave frequency ω and $B_{res} = m_e c\omega/e$ is the magnetic field corresponding to the electron cyclotron resonance.

The coefficient of the O-X mode conversion can be defined as the fraction of the electromagnetic energy flux tunneling through the evanescent region. It was shown, that for the chosen geometry, provided that length of the beam trace is less than the characteristic diffraction length, the coupling efficiency $\langle T \rangle$, averaged over the ensemble of fluctuations, can be calculated as

$$\langle T \rangle = \int_{-\infty}^{\infty} G(y,y')\tilde{A}(y)\tilde{A}^*(y')\Phi(y-y')\,dy\,dy', \quad (1)$$

Here G is the known Green function [3] defining two-dimensional structure of the mode coupling region, $\tilde{A}(y)$ is the deterministic distribution of the complex wave amplitude and function $\Phi(\Delta y)$ takes into account the additional random phase modulation $\phi(y)$ and is defined as follows:

$$\Phi(y,y') = \exp(-\langle \phi^2 \rangle + \langle \phi(y)\phi(y') \rangle). \quad (2)$$

For the small-angle scattering, the random phase is related to the density fluctuations δn_e as

$$\phi(x,y) = \int_{x_0}^{x} \beta(x')\,\delta n_e(x')\,dx', \quad \beta = \frac{\partial k_\perp}{\partial n_e},$$

where β determines the dependence of the component of the wave vector transverse to the magnetic field on the plasma density.

As a result, the phase correlation function is related to the density correlation function as follows:

$$\langle \phi(x,y)\phi(x,y') \rangle = \int_{x_0}^{x}\int_{x_0}^{x} dx'\,dx'' \times \quad (3)$$
$$\times \beta(x')\,\beta(x'')\langle \delta n_e(x',y)\,\delta n_e(x'',y') \rangle.$$

Using this equation we meet some uncertainty related to the particular choice of the upper limits of integration. In a simplest theory we may just assume $x = 0$, however, in the neat vicinity of the transformation region, $x \to 0$, the assumption about small-angle scattering may become inapplicable as diminishing values of $k_\perp$ result in a saturation of the spectral broadening for $x > x_m$ [2]. To estimate this effect, one may assume that the diffusive broadening stops when the spectral width $\langle \delta k^2 \rangle$ becomes comparable with the solution $k_\perp^2$ of the dispersion relation for the O wave propagating in the unperturbed plasma. Thus, the upper limits $x = x_m$ in (3) are defined as a negative root of the following equation

$$\left\langle \delta k^2(x_m)\right\rangle = k_\perp^2(x_m), \quad x_m < 0. \tag{4}$$

Comparison with the full-wave calculations

To verify our analytical theory, we carried out a direct numerical simulation of the tunneling of the incident O wave through the two-dimensionally inhomogeneous layer of a fluctuating overdense plasma. Maxwell equations were solved with the full-wave finite-difference time-domain code IPF-FDMC [4].

According to our basic model, the magnetic field and plasma density is assumed to vary in (x, y)-plane. Their regular profiles inside the simulation domain $[-x_0, x_0]\times[-y_0, y_0]$ is selected to reproduce the two-dimensional model shown in Fig. 1:

$$n_e/n_{cr} = \left(1 + \tanh\frac{x\cos\alpha - y\sin\alpha}{L_n}\right)\tanh\frac{x+x_0}{L_n/3},$$

$$B/B_{res} = Y - 2(Y+1)y\sin\alpha/L_n.$$

The distribution of density fluctuations is calculated as a sum over a discrete set of harmonics,

$$\delta n_e/n_{cr} = A\exp(-(x-x_{\delta n})^2/(2l_{\delta n}^2))\times$$

$$\sum_{\kappa_x,\kappa_y}\exp\left(-\frac{\kappa_x^2+\kappa_y^2}{8}\lambda^2\right)\cos\left(\kappa_x x+\kappa_y y+\xi(\kappa_x,\kappa_y)\right)$$

Here A is the normalization factor; $l_{\delta n}$ and $x_{\delta n}$ characterize size and position of the fluctuating area. For each realization, a sum over 200×100 harmonics with random phases $\xi(\kappa_x, \kappa_y)$ is generated, then the results of electromagnetic simulations are averaged over 100 realizations.

To demonstrate the effect of fluctuations, we fix the width of the incident beam and study the dependence of the averaged O-X coupling efficiency on the width $l_{\delta n}$ of the region with density perturbations, see Fig 2. Here we compare the results of numerical simulations of $\langle T\rangle$ with the analytical calculations based on equations (1)-(4) for both the one-dimensional (a) and two-dimensional (b) geometries.

Solid lines correspond to the theory with saturation of the small-angle scattering being taken into account and dotted lines correspond the simplest theory without saturation, i.e. in (3) we take $x = x_m$ in the first case, and $x = 0$ in the second case. Both theories coincide to each other and to the numerical modeling when the density fluctuations are localized far enough from the mode-coupling region $l_{\delta n} \ll |x_{\delta n}|$. When fluctuations are present inside the coupling region $l_{\delta n} \sim |x_{\delta n}|$, the saturation results in ~10% variation of $\langle T\rangle$. Note also, for the chosen beam size in two-dimensional geometry, the O-X coupling efficiency is higher than those for the same beam in one-dimensional case.

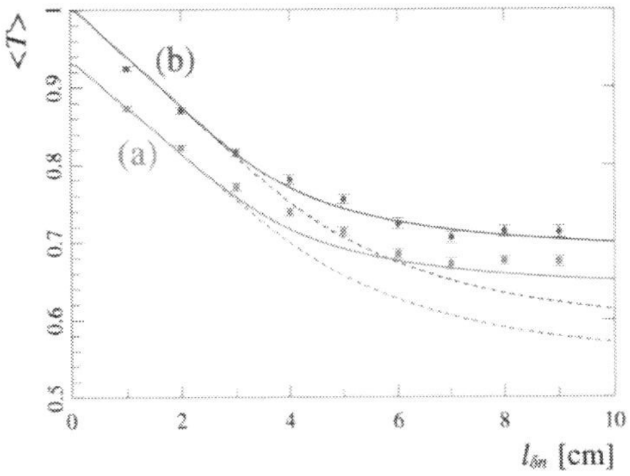

Fig. 2. The O-X mode coupling efficiency $\langle T\rangle$ of the Gaussian beam with the fixed width $a_y = 3$ cm depending on the size $l_{\delta n}$ of the region occupied by fluctuations for (a) $\alpha = 0$ and (b) $\alpha = 0.05$. Lines show predictions of the analytical theory with (solid lines) and without (dotted lines) the saturation of small-angle scattering being taken into account. Points correspond to full-wave numerical simulations; the errorbars of the numerical results are due to statistical dispersion after averaging over 100 random realizations. Other parameters of the model are: $k_0 = 12cm^{-1}$, $\delta n_e/n_{cr} = 3\%$, $x_{\delta n} = -6$ cm, $\lambda = 0.5$ cm, $L_\nabla = 0.7$ cm, $Y = 0.8$, $L_n = 10$ cm, $x_0 = 10$ cm, $y_0 = 16$ cm.

Summary

We considered the effect of a random phase modulation induced by density fluctuations along the beam propagation path on the O-X mode coupling efficiency in a two-dimensional geometry. In this case, the averaged coupling efficiency could be expressed in a universal way as a convolution of deterministic functions characterizing the geometry of the scattering region, the regular part of the beam shape, and the correlation function of the distribution of the random phase, see (1). The analytical results are verified with a full-wave code. We find a relatively good agreement of both approaches, and demonstrate the role of two physical effects related to non-one-dimensionality of the non-perturbed plasma and stopping of the stochastic spectral broadening of the wave beam near the O-X mode coupling region. Both effects modify the coupling efficiency by about 10% in a parameter range typical of magnetic fusion experiment and should be taken into account.

Acknowledgements. The work was done under Russian state assignment (projects Nos. 0035-2014-0027, 0035-2018-0003) and was supported by RFBR (pr. 18-42-520069).

References

1. *Laqua, H. P.* // Plasma Phys. Control. Fusion 2007. V. 49. No 4. P. R1.
2. *Shalashov, A. G. and Gospodchikov, E. D.* // Plasma Phys. Control. Fusion 2014. V. 56. P. 125011.
3. *Khusainov, T. A., Shalashov, A. G. and Gospodchikov, E. D.* // Plasma Phys. Rep. 2018. V. 44, No 5. P. 484.
4 *Köhn, A. et al* // Plasma Phys. Control. Fusion 2008. V. 50. P. 085018.

Overview of W7-X ECRH Results in OP1.2a

H.P. Laqua[1], J. Baldzuhn[1], H. Braune[1], S. Bozhenkov[1], K.J. Brunner[1], Ye.O. Kazakov[2], S.Marsen[1], D. Moseev[1], T.Stange[1], R.C. Wolf[1], M. Zanini[1] and Wendelstein7-X Team

[1] *Max-Planck-Institute for Plasma Physics, Greifswald, Germany*
[2] *Laboratory for Plasma Physics, LPP-ERM/KMS, Brussels, Belgium*

The stellarator Wendelstein7-X was exclusively heated by ECRH in the operation phase OP1.2a. The ECRH system consists of 10 gyrotrons, with an output power of 0.6-1.0 MW each, a quasi optical transmission line and flexible microwave launcher inside the plasma vessel [1]. The over-all transmission efficiency was estimated to be ~ 94%. The ECRH system is commissioned for 1800 s operation at full power. It already demonstrated all requirements, which are necessary for a high-performance steady state operation at reactor relevant parameters at W7-X. The 140 GHz ECRH uses the second harmonic resonance at 2.5T with an absorbed power of up to 7 MW. Besides the reliable plasma start-up und routine ECRH wall conditioning, stationary discharges up to 30 s have been achieved, which were only limited by the maximum test divertor energy load. Furthermore, the novel remote steering launcher, an important concept for the future fusion reactor, was tested for the first time for plasma heating and current drive. The long discharges were used to demonstrate current control and bootstrap current compensation by ECCD. Localized ECCD strongly changed the rotation transform iota and thus the confinement and stability. In particular by dedicated ECCD strong MHD activity with repetitive central temperature collapse could be driven, which in the worst case lead to a total loss of the plasma confinement. In combination with pellet injection (PI), highest performance with a plasma energy above 1 MJ has been achieved with the X2-mode ECRH at a plasmas density of 0.8 10^{20} m^{-3} and an ion temperature of 3.8 keV, which demonstrates the already good collisional coupling between electrons and ions at that density. Even higher densities have been achieved using the O2-mode combined with PI as shown in Fig.1. Here high-performance plasmas at densities above the X2-cutoff and up to 1.4 10^{20} m^{-3} have been demonstrated, which is already close to the envisaged future steady state high-performance plasma scenario. Achieving efficient plasma heating with the O2 mode requires establishing a sufficiently dense and hot target plasma for the O2 mode to take over. This was accomplished using the X2-mode for the plasma start-up, and then during the initial two seconds of plasma operation the polarization was changed to O2 before the start of pellet fuelling. The O2-mode heating efficiency of > 90% was achieved with the help of individually shaped reflector tiles for the 10 ECRH beams enabling at least 3 passes through the EC-resonance in the plasma center. The reliable and efficient operation was guaranteed by a set of ECRH protective diagnostics, which were specially established for the ECRH requirements. In particular the ECRH stray radiation measurement turned out to be a reliable interlock signal to prohibit ECRH operation with insufficient plasma absorption. Furthermore the position and launch direction of each beam was measured by infrared cameras observing its thermal footprint at the heat shield tiles opposite of the antennas. Finally selected 140 GHz beams was also used as a source for the collective Thomson scattering (CTS), which was also commissioned [2]. Here a local ion temperature measurement was demonstrated (see Fig.2). For the next operational campaign op1.2b new ECRH scenarios have been prepared. In particular the operation at a reduced magnetic field strength of 1.75 T requires the use of the third harmonic x-mode (X3). Here ray-tracing calculation predicted sufficiently high absorption for plasma build-up and sustain if the electron temperature is above 200 eV. The plasma start-up with X3 needs probably an assistance by another heating method, like X2 ECRH at 105 GHz or ICRH. It is also planed to test a new method to excite waves at the ion cyclotron (IC) frequency with ECRH. The ECRH power will be modulated with the IC-frequency using the beat wave of two crossing ECRH beams. For the conversion into IC-oscillation the magnetic field strength variation along the magnetic axis will be used. The electron perpendicular magnetic momentum **M** will be increased by the EC-interaction and the grad **B·M** force will drive the oscillation.

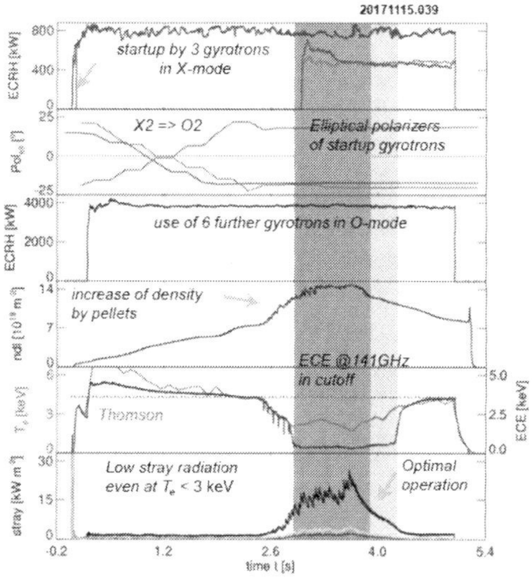

https://doi.org/10.1051/epjconf/201818701011

Fig. 1. High density plasma scenario with O2-ECRH and pellet injection. The second row shows the chance of polarizer position for the 3 "start-up" gyrotron beams, when their polarisation was switched from X2 to O2.

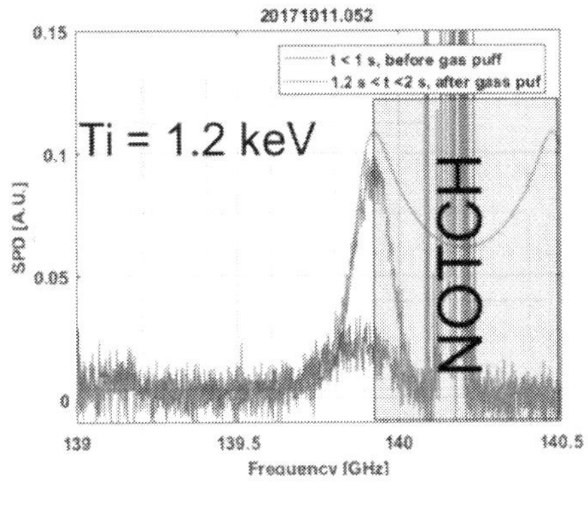

Fig. 2. CTS spectrum at low density (bleu) and medium density (red) in a Helium plasma.

1. *Erckmann V. et al.(2007)* Electron Cyclotron Heating for W7-X: *Physics and Technology, Fusion Science and Technology, 52:2, 291-312, DOI: 10.13182/FST07-A1508*

2. *Moseev D. et al.* Collective Thomson Scattering Diagnostic at Wendelstein7-X *submitted to review of scientific instruments*

EPJ Web of Conferences **187**, 01012 (2018) https://doi.org/10.1051/epjconf/201818701012
30th Joint Russian-German Meeting on ECRH and Gyrotrons

Using Fullwave Simulations to Understand the Turbulent Wavenumber Spectrum Measured by Doppler Reflectometry

C. Lechte[1], G. D. Conway[2], T. Görler[2], T. Happel[2], C. Tröster-Schmid[2], the ASDEX Upgrade Team

[1]IGVP, University of Stuttgart, Germany, carsten.lechte@igvp.uni-stuttgart.de
[2]Max-Planck-Institut für Plasmaphysik, Boltzmannstr. 2, 85748 Garching, Germany

Doppler reflectometry is an important microwave diagnostic for turbulent fusion plasma. A low power probing beam between 40 and 100 GHz in O or X mode polarisation is scattered at certain density fluctuation wavenumbers, depending on frequency and angle between beam and density gradient. This way, a radially localised wavenumber resolved density fluctuation spectrum can be measured. However, the correspondence between fluctuation power and scattered microwave power is strongly non-linear and is investigated using fullwave simulations, with density fluctuations supplied by a plasma turbulence code. Earlier results for X mode polarisation have shown saturation of the spectral power to occur at high fluctuation amplitudes, an enhancement at intermediate amplitudes, and linear behaviour at low amplitudes. Together, these effects strongly modify the shape of the wavenumber spectrum [1].

Experimental data in X and O mode polarisation and accompanying plasma fluctuation data from the GENE code have been used to simulate the Doppler reflectometer response with the fullwave code IPF-FD3D [2].

The gyrokinetic plasma turbulence code GENE is a physically comprehensive, hyperscalable Vlasov code that is used with the same ASDEX Upgrade equilibrium and plasma profiles as the experiment and the fullwave simulations. A flux tube surrounding the radial position of the microwave cutoff was simulated. The input parameters were adjusted slightly until the resultant turbulent fluxes matched the ones observed in experiment. The output density fluctuation fields are used as inputs in the fullwave simulations.

IPF-FD3D is a finite difference time domain code for cold plasmas. Simulations were done in 2D for both polarisations in the perpendicular plane for 512 different turbulence time slices. The turbulence is considered frozen in time while the wave is propagating. The antenna angles and frequencies were adjusted to probe the same ρ_{pol} as the experiment. The setup is shown in Fig. 1.

Comparison with experimental data showed a good agreement [2], with the O mode spectrum corresponding more closely to the GENE density spectrum. Furthermore, the fullwave simulation data in Fig. 2 shows the big difference between O and X mode polarisation, which were both probing the same fluctuations. The spectrum derived from O mode scattering looks similar to the real spectrum of the density fluctuations, albeit with power law exponent -8.7 instead of -6.7. The X mode spectrum is much shallower at the nominal fluctuation amplitude taken directly from GENE. In addition, the kink or knee of the spectrum is moved to larger wavenumbers compared to O mode and GENE. The knee should be the same, as it is determined by the instability that is driving the turbulence.

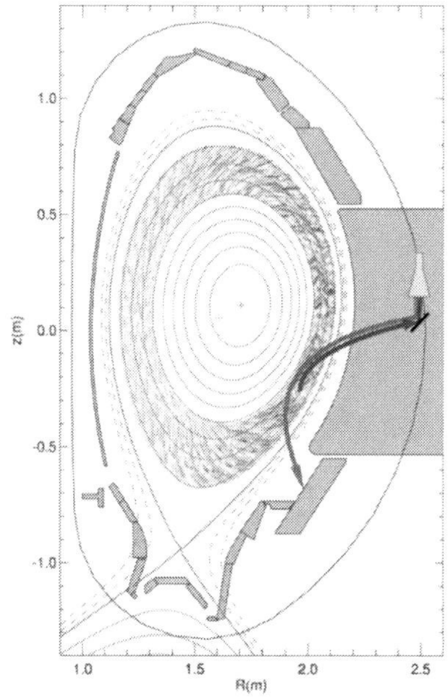

Fig. 1. Poloidal geometry of ASDEX Upgrade with Doppler reflectometer beams and superimposed density fluctuation field from GENE

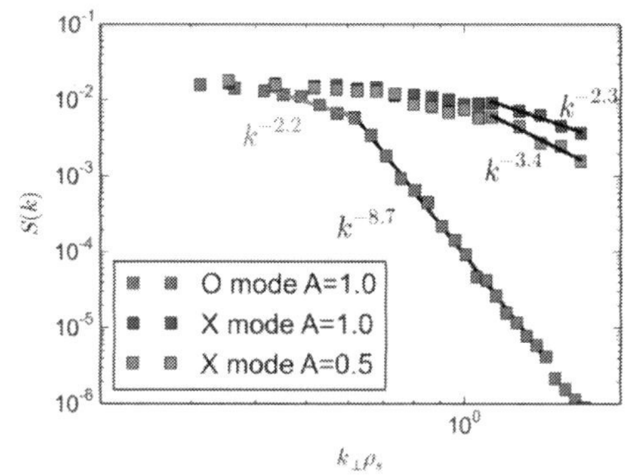

Fig. 2. Wavenumber spectra from fullwave Doppler reflectometry simulations for both X and O mode polarisation with fitted spectral indices.

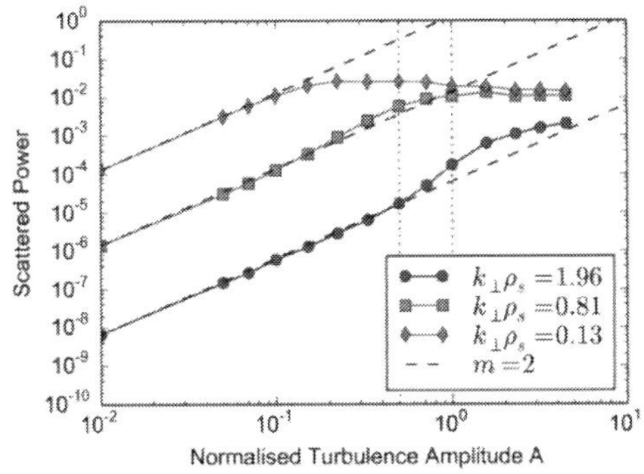

Fig. 3. The scattered power plotted over the density fluctuation amplitude for 3 exemplary wavenumbers for X mode polarisation

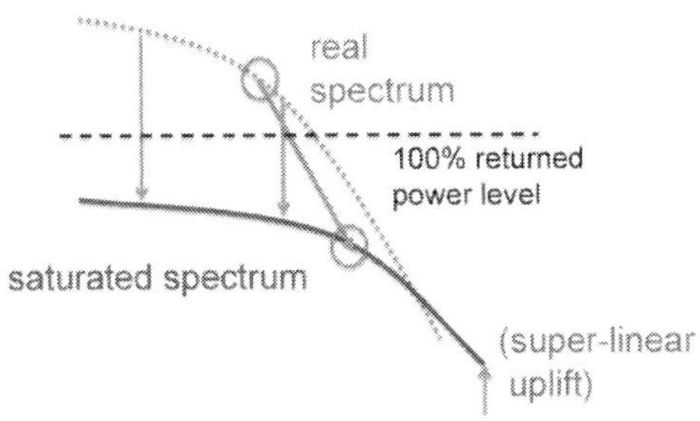

Fig. 4. Schematic wavenumber spectra illustrating the observed deformation of the spectrum caused by the sensitivity shifts shown in Fig. 3

A factor of 0.5 was necessary to bring the simulated Doppler spectrum in line with the experimental one. This is justified because of some uncertainties in the temperature profiles.

Scaling down the fluctuation amplitude reveals steeper spectra, which also do not have the saturation effect at low wavenumbers. This means that for X mode, the density fluctuations are so large that the spectrum is non-linearly saturated, while for O mode, it is still in the linear regime. This was simulated in more detail with simple isotropic turbulence, and is shown in Fig. 3.

The scan of fluctuation amplitudes (see Fig. 3) shows that a non-linear saturation sets in at a scattered power of 0.01 a.u. Due to the high level of low-$k_\perp$ fluctuations, this part of the spectrum is saturated, while the high-$k_\perp$ fluctuations follow a linear relation, but with a non-linear enhancement. This had already been reported in [3]. The turbulence amplitude A is normalised to the GENE results.

The overall effect on the spectrum for X mode Doppler reflectometry is illustrated in Fig. 4. The shape of the density fluctuation spectrum (labeled 'real spectrum') is strongly decreased at the low-$k_\perp$-high-amplitude regime because of non-linear saturation. At the high-$k_\perp$-low-amplitude end of the spectrum, we see the non-linear enhancement. The overall effect is a much shallower spectrum with the knee position apparently shifted to higher $k_\perp$ (indicated by the green arrow).

In O mode polarisation, the non-linear saturation effect is much less pronounced, resulting in a much steeper spectrum. This is attributed to the lower frequencies that are necessary to probe the same wavenumbers in O mode.

In summary, the discrepancy between the GENE density fluctuation spectra and the Doppler reflectometry wavenumber spectra can be explained by the non-linear saturation and enhancement properties of the diagnostic itself. This also helps validate the turbulence simulations.

Acknowledgements

The simulations were performed on the HPC-FF computer cluster in Jülich, Germany; on the BOB Linux cluster at RZG; on hornet and hazelhen at HLRS Stuttgart; on bwUniCluster funded by the State and the universities of Baden-Württemberg, Germany, within the framework programme bwHPC. This work was also partly performed within the framework of the Helmholtz Virtual Institute on Plasma Dynamical Processes and Turbulence Studies using Advanced Microwave Diagnostics.

References

[1] *C. Lechte et al.* "X mode Doppler Reflectometry k-spectral measurements in ASDEX Upgrade: Experiments and simulations" *Plasma Phys. Contr. Fus.* **59** (7) 2017 doi:10.1088/1361-6587/aa6fe7

[2] *T. Happel et al.* "Comparison of detailed experimental wavenumber spectra with gyrokinetic simulation aided by two-dimensional full-wave simulations" *Plasma Phys. Contr. Fus.* **59** (5) 2017 doi:10.1088/1361-6587/aa645b

[3] *J. R. Pinzón et al.* „Enhanced Doppler reflectometry power response: physical optics and 2D full wave modelling" *Plasma Phys. Contr. Fus.* **59** 2017, doi:10.1088/1361-6587/aa543c

Wave Beams of Short Radio Pulses Generated by Gigawatt Microwave Sources

A. V. Palitsin, M. B. Goykhman, A. V. Gromov, N. F. Kovalev

Institute of Applied Physics of the Russian Academy of Sciences, Nizhny Novgorod, Russia, pal@appl.sci-nnov.ru

A rising interest to the generation of short pulses is observed lately [1-3]. In relativistic backward wave oscillators (BWO) operating in superradiance regime the peak power of output radiation may exceed the power of electron beam [1,2]. Such BWOs are effectively synchronized by the leading edge of accelerating voltage pulse so that the multi-channel systems can be built [1,2]. Radiation patterns of these systems consist of overlapping wave beams. The structure of each wave beam is usually close to the fundamental Gaussian beam with linear polarization. Dimensions of the output window of each channel exceed wavelength λ because of microwave breakdown on smaller apertures. It means that the distance between channels d is greater (typically much greater) than wavelength

$$d >> \lambda \tag{1}$$

and far field radiation pattern has many maximums.

In the report we analyze theoretically the peculiarities of such multi-channel systems in case of radiating the phase synchronized short pulses. Superposition of 2D scalar Gaussian beams (see Fig. 1) has been considered:

$$E = \mathrm{Re}\sum_{n=1}^{N}\left(\sqrt{\frac{-ika^2}{z-ika}}\exp\left[ik\frac{(X-X_n)^2}{2(z-ika^2)}\right]\times \right. \tag{2}$$
$$\left. \times\exp(i2\pi ft-ikz)\right).$$

In (2): E is normalized (dimensionless) electric field; $k = 2\pi/\lambda = 2\pi f/c$ is a wavenumber; f is a frequency; c is a velocity of light; x and z are transverse and longitudinal coordinates respectively; X_n is a coordinate of the Gaussian beam with index n; and a defines the transverse dimension of the beam waist.

Patterns of energy flux density P in far field region have been considered for radiation of:

i. short pulses with the central frequency f_0 and Gaussian envelope

$$E_p = \mathrm{Re}\left(\exp\left[-\frac{t^2}{2\tau^2}\right]\exp(i2\pi f_0 t)\right), \tag{3}$$

where 2τ is a pulse duration (see Fig. 2).

ii. CW signals with the same frequency f_0.

For the short pulses the radiation pattern $P_p(X)$ is defined as a peak value of energy flux density in the pulse transmitted through the point X in the far field cross section $z = L$. For CW signals the radiation pattern $P_f(X)$ is equal to energy flux density in CW signal at the same cross section $(X, z = L)$.

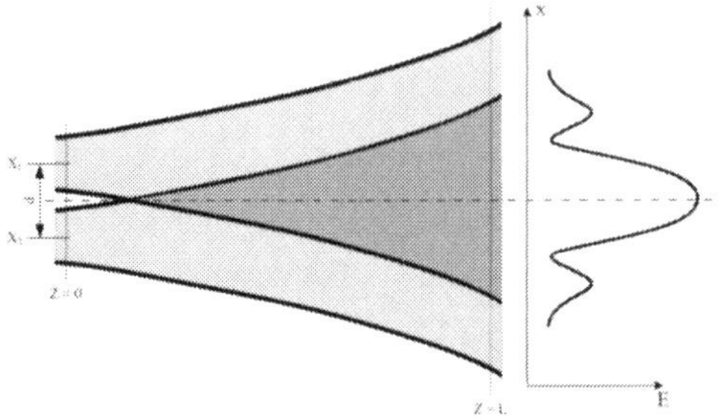

Fig. 1. Superposition of Gaussian beams and distribution of electric field at coordinate $z = L$ in far field region (shown schematically).

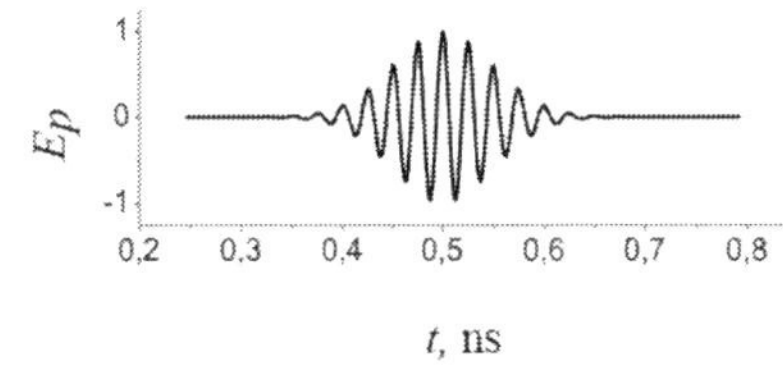

Fig. 2. Short pulse with the central frequency $f_0 = 40$ GHz (as an example) and Gaussian envelope.

Summation of two Gaussian beams resulted in small differences of radiation patterns for CW and short pulses signals as it is shown in Fig. 3. However, all maxima except the main one with highest amplitude are slightly reduced. In case of superposition of 8 Gaussian beams the suppression of side maxima becomes essential (see Fig. 4). Shapes of pulses outside the main maximum are distorted as it is shown in Fig. 5b.

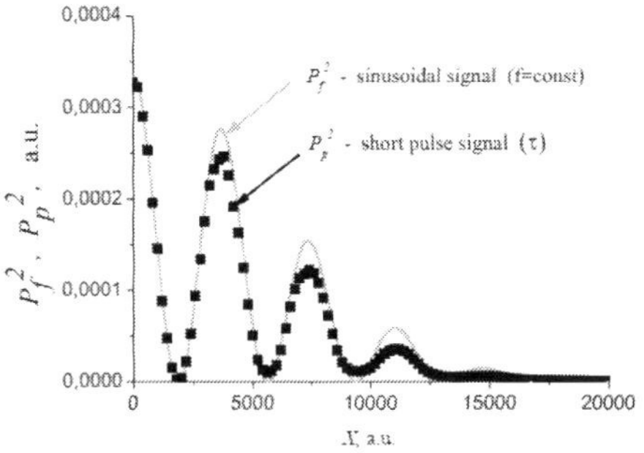

Fig. 3. Radiation patterns of CW (solid gray line) and short pulse (black dots) signals simulated for superposition of two Gaussian beams.

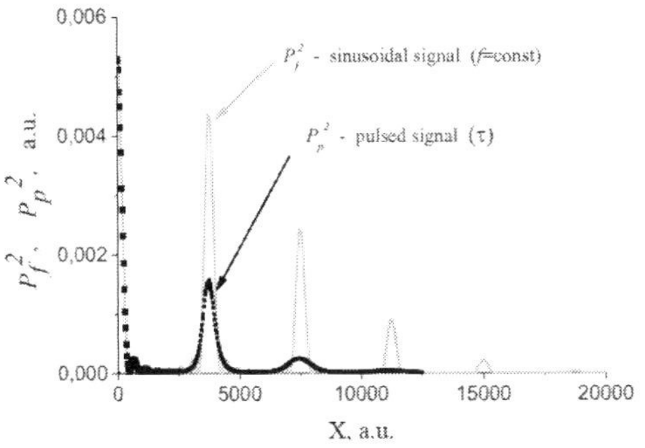

Fig. 4. Radiation patterns of CW (solid gray line) and short pulse (black dots) signals simulated for superposition of 8 Gaussian beams.

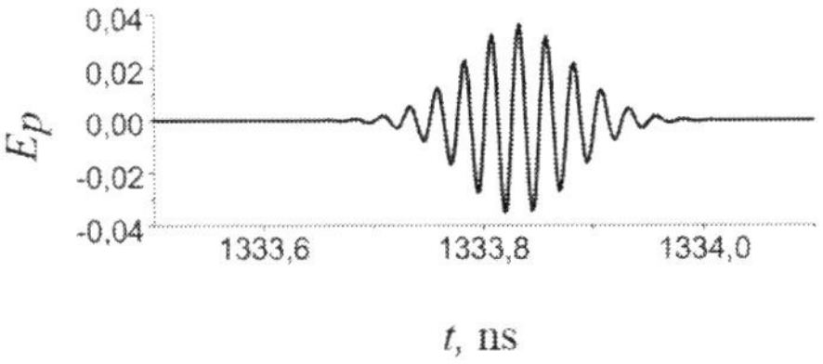

(a)

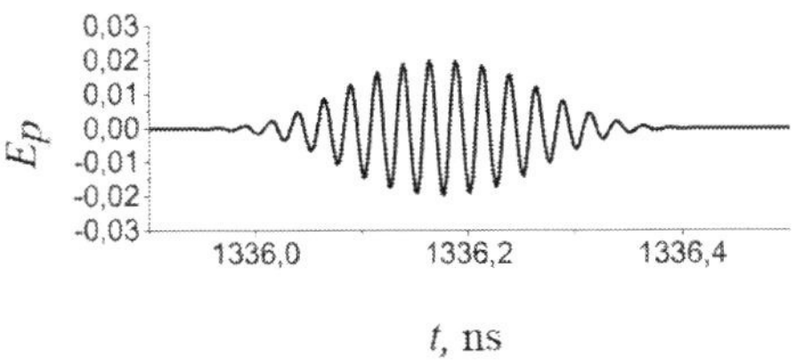

(b)

Fig. 5. Pulses in two points of far field cross section $z=L$. **a** - in the main maximum $X=0$ and **b** - in the first side maximum $X\neq 0$.

The reason of such distortion and accordingly the suppression of side maxima in radiation pattern is a delay of pulses transmitted in different wave beams and observed at the same point $(X\neq 0, z=L)$ in the far field region. It is appropriate to introduce a novel parameter for such multi-beam radiation systems

$$L_{corr}=\frac{2\tau\cdot c}{\sin(\theta)} \tag{4}$$

a correlation length. Here 2τ is the pulse duration and θ is an angle between z axis and the direction to the point of observation (see Fig. 6). For large enough numbers N of wave beams at any point of observation only pulses in wave beams with X_n from the range

$$\Delta X_n \leq L_{corr} \tag{5}$$

overlap and can be coherently summed. The number N_{corr} of such beams with overlapping pulses can be named as correlation number.

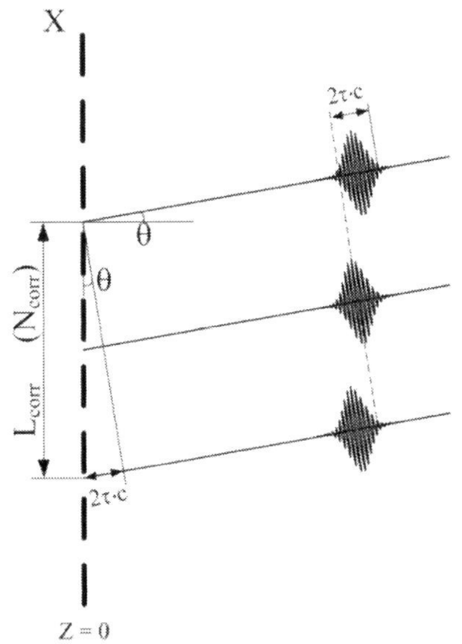

Fig. 6. Explanation for the correlation length L_{corr} and correlation number N_{corr}.

For the angle $\theta=0$ pulses with any durations are overlapping because $L_{corr}(\theta=0)=\infty$. It is a reason for invariability of main maximum in radiation pattern which is proportional to the square of the total number of wave beams N^2. For any side maximum $(\theta\neq 0)$ the correlation length is finite (if the pulse duration is finite) and the amplitude of side maxima is proportional to $N_{corr}^{\ 2}$ as electric fields only from N_{corr} wave beams are coherently summed. Consequently, for large enough numbers N of the beams the side maxima level of radiation pattern for short pulse signals can be suppressed as low as

$$\frac{N_{corr}^{\ 2}}{N^2}\to\frac{1}{N^2} \tag{6}$$

as it is compared to side maxima of CW radiation pattern.

References

1. *A. Sharypov, A. A. El'chaninov, G. A. Mesyats, M. S.Pedos, I. V. Romancheko et al.* Coherent summation of Ka-band microwave beams produced by subgigawatt superradiance backward wave oscillators // Appl. Phys. Lett. 103. 134103 (2013).

2. *N.S. Ginzburg, A.W. Cross, A.A. Golovanov, et al.* Coherent summation of emission from relativistic cherenkov sources as a way of production of extremely high-intensity microwave pulses // IEEE Trans. on Plasma Science. 2016. V. 44. N 4. P. 377-385.

3. *N.S.Ginzburg, R.M.Rozental, A.S.Sergeev, A.E.Fedotov, I.V.Zotova, V.P.Tarakanov.* Generation of rogue waves in gyrotrons operating in the regime of developed turbulence // PRL 119. 034801 (2017).

The newest ceramic materials for microwaves

V.V. Parshin[1], E. A. Serov[1], P.V. Netcvetaeva[2]

[1]Institute of Applied Physics of RAS, N.Novgorod, Russia, parsh@appl.sci-nnov.ru
[2]OJSC "Magneton Plant", St. Petersburg, Russia

The refractive index and absorption (tanδ) investigations in different kind of ceramic materials produced in Russia and Germany were made by using the open high-quality Fabry-Perot resonator in the frequency range 50-200 GHz in IAP RAS [1]. The low frequency measurements of the samples were made using cavity resonator by producer of materials [2].

It was established that the quality of these new production is the same as it was in Soviet Union and comparable with the world level. This paper presents the result of investigations.

Fig. 1. Vacuum tight foam ceramics produced by JSC "Magneton Plant" with extra low refractive index values

Table 1

	n / ε f = 9.4 GHz	tanδ [10^{-4}] f = 9.4 GHz	ρ [g/cm^3]
LK-1.5	1.22 / 1.5	≤ 5	0.48
LK-2	1.41 / 2	≤ 8	0.8
LK-2.5	1.58 / 2.5	≤ 12	1.2
LK-3	1.73 / 3	≤ 15	1.5

Low frequency parameters of vacuum tight foam ceramics are presented in Table 1. The loss tangent frequency dependencies of grades LK-2.5 and LK-3 are presented in Fig. 2. The vacuum tight foam ceramics are very interesting materials for microwave applications because it is possible to slightly change the foam content and by this way to get the needed refractive index value. But measurements are meaningless above 200 GHz due to strong losses increasing mainly due to scattering by "foam". By this reason the grades LK-1.5 and LK-2 are not for MM waves.

At the range 100 – 200 GHz the refractive index increases practically linearly from n=1.58 (9.4 GHz) to n=1.61 (170 GHz) for LK-2.5 and from n=1.73 (9.4 GHz) to n=1.763 (192GHz) for LK-3.

The main parameters of ceramics produced by JSC "Magneton Plant" (St. Petersburg) are presented in Table 2. (at 170 GHz). The loss tangent frequency dependencies of "traditional" ceramics are practically linear at MM waves and refractive indices are practically constant.

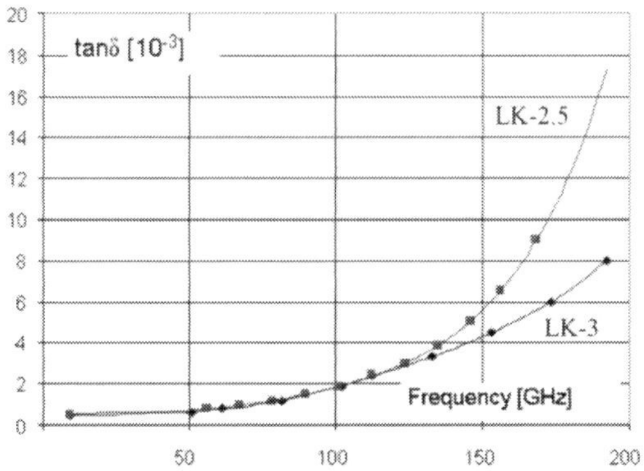

Fig. 2. The loss tangent frequency dependencies of grades LK-2.5 and LK-3

Table 2

Grade	n	tanδ [10^{-4}]	ρ [g/cm^3]
MST-7.3	2.685	8.5	3.18
MST-10	3.185	9	3.37
BK-100M	3.165	4	3.98
TK-20	4.416	50	3.78
TK-40	6.255	37	4.78
LK-2.5	1.61	9	1.2
LK-3	1.76	50	1.5
ST-3	1.707	20	1.15
ST-4	1.995	21	1.35
ST-10	3.194	36	2.00

MST grades are the ceramic materials with very stable refractive index values which are traditionally used as constructive elements in microwave technology. Materials can be made in the different form including very complicated ones.

TK grades are thermally stable microwave ceramics which are widely used to create matching and structural elements, substrates for integrated circuits, filters, dielectric resonators with increased requirements for thermostability of dielectric characteristics.

ST grades are microwave dielectrics based on polymers filled with titanium dioxide. In a large-scale production the materials of this series are the basis for the manufacture of products of complex shape with high accuracy by injection molding methods.

BK corundum ceramics are the alumina-based material with various modifying additives. It is traditionally used for manufacturing of substrates for integrated microwave circuits, substrates for high voltage stabilizers, microstrip device supports, as well as high-voltage insulators and energy output windows for medium-power generators.

The main parameters of ceramics produced by "Friatec" (Germany, http://www.friatec.de) are presented in Table 3. Fig. 3. presents the results of loss tangent measurements. It is the traditional ceramics on the base of different kind of oxides with different additives for "improving" of electromechanical properties.

Table 3

Materials (170 GHz)	n	tanδ $[10^{-4}]$	k [W/cm·K]	α $[10^{-6}$ $K^{-1}]$	E GPa
Degussit (AlN)	2.91	5.8	1.8-2	3.6	330
Frialit HP79 (Si_3N_4)	2.07	70			
Degussit Al23 HF (Al_2O_3)	3.14	2.7	0.35	7.3	380
Degussit Al23 (Al_2O_3)	3.313	3.6	0.35	8.2	350
Frialit 99.7 HF (Al_2O_3)	3.15	2.7	0.35	7.3	350
Frialit 99.7 (Al_2O_3)	3.15	6	0.35	7.3	350
Frialit FZT (ZrO_2)	3.36	9	0.25	7.5	460
Frialit FZM (ZrO_2+Mg)	5.46	40	0.03	10.4	500

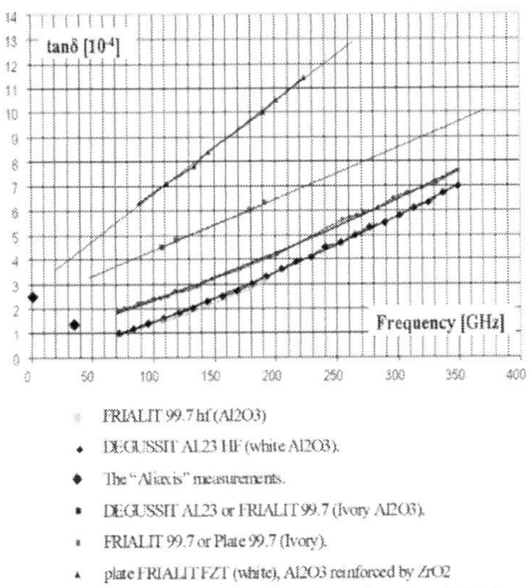

Fig. 3. Loss tangent frequency dependencies of "Friatec" samples

Conclusion

At present, the JSC "Magneton Plant" (St. Petersburg) (http://www.magneton.ru) has mastered the mass production of 35 grades of dielectric materials continuously overlapping the range of dielectric constant (ε) values from 1.5 to 120. Work is in progress to start producing the grades with the values of $\varepsilon \approx 140$, as well as to increase the thermal resistance of dielectric materials based on polymers to temperatures of 120 ÷ 250°C [1].

As a result of the research, it can be concluded that a number of materials, with low absorption, are quite suitable for use in MM and partially in the SubMM ranges for various structures interacting with electromagnetic wave.

The work was partly supported by the Russian Foundation of Basic Research and the Government of the Nizhny Novgorod region, project No. 18-42-520015.

References

1. *Parshin, V. V., Tretyakov, M. Yu., Koshelev, M. A., Serov, E.A.* Modern resonator spectroscopy at submillimeter wavelengths // IEEE Sensors Journal. 2013. V. 13, No 1. P. 18-23.

2. *Firsenkov, A. I., Kanivetc, A. J., Kasatkina, T. C., Ershova, O. M., Ivanova, L. P., Ershova, P. V.* Microwave dielectric materials produced by JSC "Magneton Plant" // Proceeding of IV-th All Russian conference Microwave Electronics and microelectronics. St.-Petersburg: ETU «LETI» publishing house , 2015. – V 2. – P. 55-59.

Synthesis of reflection gratings for advanced plasma heating scenarios

B. Plaum[1], M. Schubert[2], A. Zeitler[1], W. Kasparek[1], C. Lechte[1], J. Stober[2] and ASDEX Upgrade Team

[1]Institut für Grenzflächenverfahrenstechnik und Plasmatechnologie, Universität Stuttgart, Pfaffenwaldring 31, D-70569 Stuttgart, Germany, plaum@igvp.uni-stuttgart.de
[2]Max-Planck-Institut für Plasmaphysik, Boltzmannstr.2, D-85748 Garching, Germany

When heating fusion plasmas at higher harmonics of the electron cyclotron frequency, the absorption efficiency can be reduced. This leads to a significant transmitted beam power hitting the wall at the high field side (HFS).

To protect the wall and to use the remaining power, one method is to place a specialized reflector at the HFS, which directs the beam back into the plasma for a second heating pass [1]. The reflector needs to conform to the tiles at the HFS, which are often nonplanar. The direction of the reflected beam is chosen such that the absorption of the second heating pass is maximized and the wall area at the LFS, which is hit after the second pass, contains no sensitive components. Additional requirements are a refocusing of the beam and polarization independence of the reflection characteristics. All these conditions can only be fulfilled with a grating. The design process involves the decomposition of the 3D field problem into a series of 2D reflections of plane waves. After the 2D gratings are optimized, the final 3D grating is synthesized.

In earlier works [1], [2] a code for the optimization of such gratings was already developed and several tiles were successfully designed. This code was modernized and a number of new features were added.

The incident and reflected beams are defined in terms of astigmatic Gaussian beam parameters and the orientations of the beam axes. Together with the basic shape of the reflector (e.g. saddle-type for the HFS of a tokamak) local k-vectors $\mathbf{k}_{in}$ and $\mathbf{k}_{out}$ can be calculated for every point on the mirror. This 3D problem can now be described by an ordinary reflection parallel to the groove and the diffraction of a plane wave at an infinite grating perpendicular to the groove.

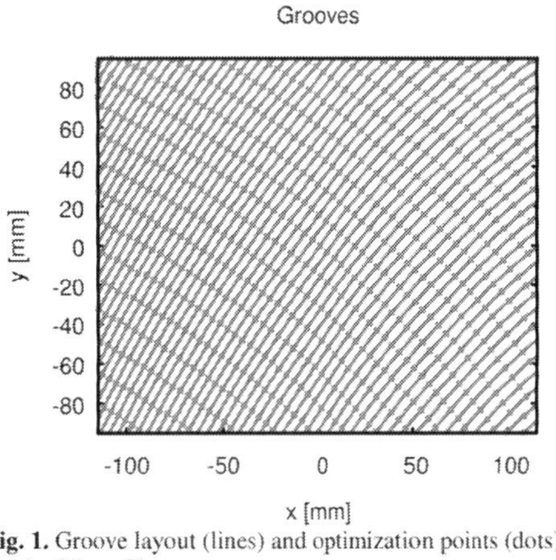

Fig. 1. Groove layout (lines) and optimization points (dots) for the 2D profiles

The only parameters describing the 2D diffraction are the beam angles Θ_{in} and Θ_{out} and the diffraction order n. Since the direction of the grooves can be calculated from $\mathbf{k}_{in}$, $\mathbf{k}_{out}$ and the normal vector, the whole groove layout can be obtained by a simple Runge Kutta integration of the grating vectors. Along the grooves, we define the points, for which a 2D profile will be optimized (see Fig. 1).

The goal of the optimization is to maximize the reflected power in the desired order for both polarizations and to minimize the phase shift between the polarizations. The minimum phase shift is necessary to preserve an incident elliptical polarization, which is required for current-drive scenarios.

The profile of a single groove is described by Fourier coefficients, where the maximum order can be obtained with the sampling theorem from the ratio of the groove period to the wavelength.

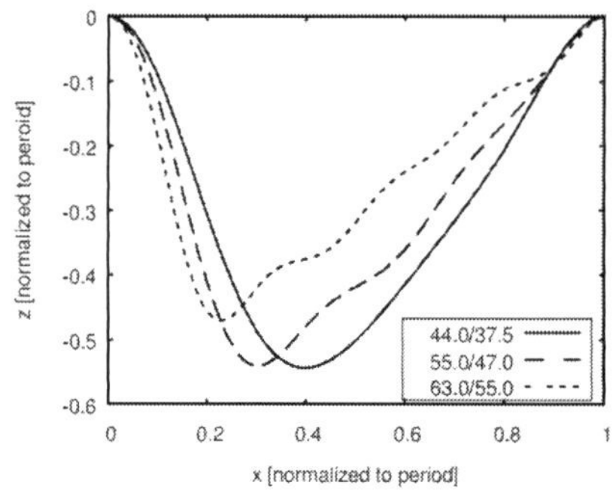

Fig. 2. Optimized 2D profiles for different combinations of Θ_{in} and Θ_{out}

Fig. 2 shows optimized 2D profiles for different combinations of Θ_{in} and Θ_{out}. The grooves are shifted such, that the highest points are at the interval boundaries. This ensures a smooth transition between adjacent grooves.

The final 3D grating is then synthesized by interpolation, Fig. 3 shows the scheme. The cross is the point (x,y), for which the local z-coordinate is calculated. The first step is to identify the groove. Then, the mirror coordinates are transformed into groove local coordinates (u,v). The Fourier coefficients are then obtained from the nearest optimized profiles by a simple 1D interpolation in the direction of the v-coordinate.

Fig. 4 shows the final interpolated grating of a reflector, which was designed for the ECRH-2 system of ASDEX-Upgrade [3]. It resembles the groove layout from Fig. 1 and shows also the saddle type shape of the reflector. The reflector was manufactured from a graphite blank using an NC milling machine (See Fig. 5).

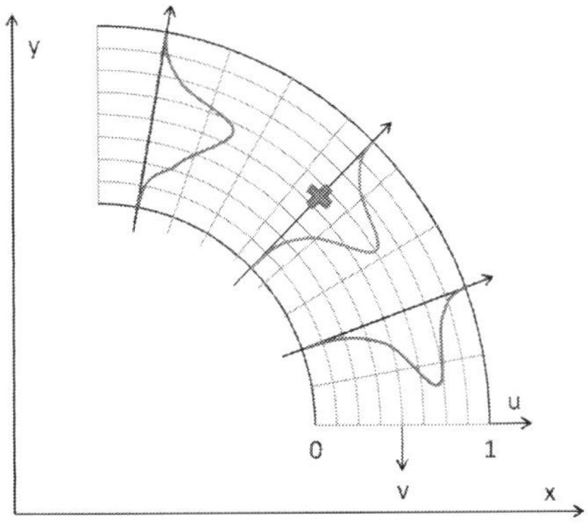

Fig. 3. Interpolation scheme for the final 3D grating. Note that in general the groove width is not constant

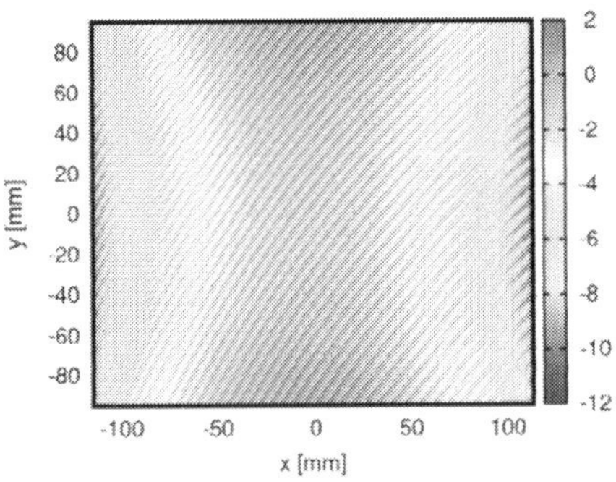

Fig. 4. Interpolated grating

Fig. 5. Reflector in the milling machine

For an experimental characterization the reflector was mounted on a goniometer and illuminated by a horn antenna with a Gaussian beam. The reflected field was measured with an x-y scanner and a vector network analyzer. The measured intensity and phase are shown in Figs. 6 and 7. One can see a nicely focused beam and an astigmatic phase similar to the design value.

The reflector was coated with a thin tungsten layer (13–15 μm) to conform to the ASDEX-Upgrade requirements and is already installed in the tokamak. High power tests will be performed during the next experimental campaign.

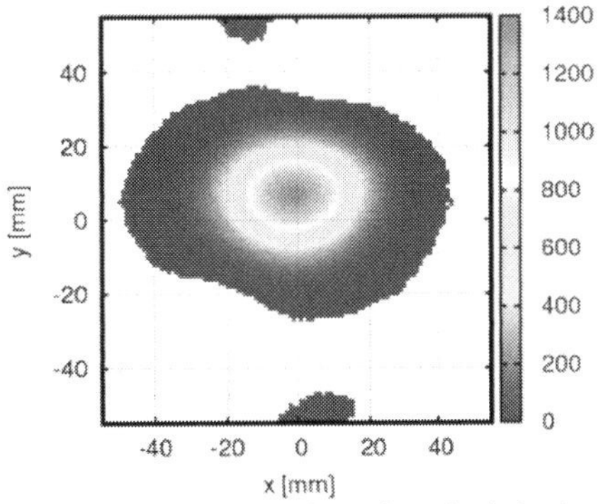

Fig. 6. Measured intensity for horizontal polarization

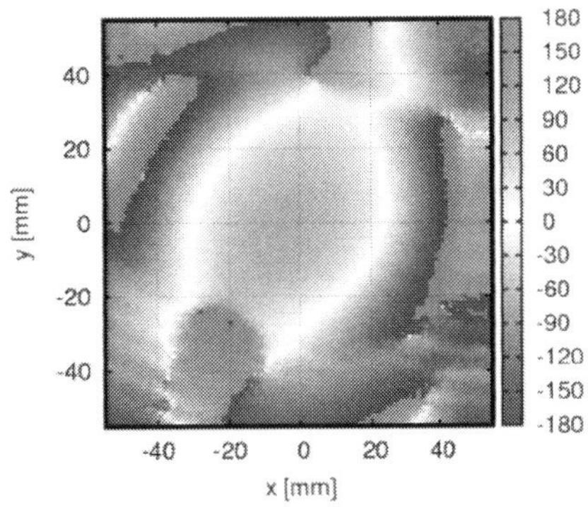

Fig. 7. Measured phase for horizontal polarization

References

1. *H. Höhnle, J. Stober, A. Herrmann, W. Kasparek, F. Leuterer, F. Monaco, R. Neu, D. Schmid-Lorch, H. Schütz, J. Schweinzer, U. Stroth, D. Wagner, S. Vorbrugg, E. Wolfrum and the ASDEX Upgrade Team*, Extension of the ECRH operational space with O2 and X3 heating schemes to control tungsten accumulation in ASDEX Upgrade, Nucl. Fusion, **51**, 083013 (2011)

2. *O. Mangold*, Effiziente holographische Gitterspiegel für die Elektronzyklotronresonanzheizung von Fusionsplasmen, Doctoral Thesis, Universität Stuttgart, 2009

3. *D. Wagner, J. Stober, F. Leuterer, F. Monaco, S. Müller, M. Münich, C. J. Rapson, M. Reich, M. Schubert, H. Schütz, W. Treutterer, H. Zohm, M. Thumm, T. Scherer, A. Meier, G. Gantenbein, J. Jelonnek, W. Kasparek, C. Lechte, B. Plaum, T. Goodman, A.G. Litvak, G.G. Denisov, A. Chirkov, V. Zapevalov, V. Malygin, L.G. Popov, V. O. Nichiporenko, V. E. Myasnikov, E. M. Tai, E.A. Solyanova, S.A. Malygin, ASDEX Upgrade Team*, Status, Operation, and Extension of the ECRH System at ASDEX Upgrade, Journal of Infrared, Millimeter, and Terahertz Waves, 37, 45-54 (2016)

Status of the gyrotron complex for ITER: composition of the complex, manufacturing, obtained parameters, delivery conditions

L. Popov[2], M. Agapova[2], Yu. Belov[2], A. Chirkov[1], G. Denisov[1], A. Eremeev[1], A. Gnedenkov[2], V. Ilin[2], A. Kuzmin[2], A. Litvak[1], A. Lyubimov[2], V. Malygin[1], V. Miasnikov[2], V. Nichiporenko[2], E. Sokolov[1], E. Soluyanova[2], E. Tai[2], S. Usachev[2], V. Usov[1], V. Zapevalov[1]

[1] Institute of Applied Physics RAS, Nizhny Novgorod, Russia
[2] Gycom Ltd., Nizhny Novgorod, Russia

IAP RAS in cooperation with Gycom Ltd. prolongs manufacturing of gyrotron systems (RF power sources) for ITER under contract #1770641334818000085O/45-320 of the 27.04.2018 with Russian DA "ITER Center". Serial production had started after approval of Final design review in December 2016 following by approval of Manufacturing readiness review (MRR) in April 2017. MRR confirmed that Gycom has all necessary facilities for full cycle of gyrotron production: machining, brazing, assembling, exhausting and testing.

Since that time two gyrotron systems were completed and successfully put through acceptance test meeting ITER specifications (see[1, 2]). Additional 1MW/170GHz/1000s gyrotron identical to serial one was produced by Gycom under request of F4E and shipped to Swiss Plasma Center facility in Lausanne after acceptance test carried out at Gycom with participation of F4E representatives in April 2017.

Gyrotron system components will be located on 2nd and 3rd floors of ITER building B15 as schematically shows this picture.

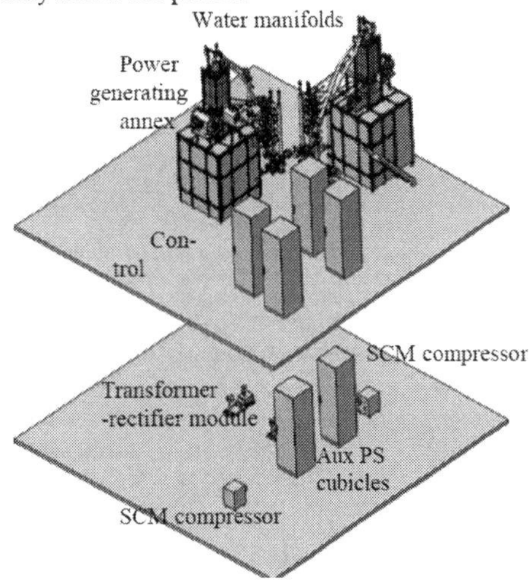

Fig. 1 Schematic view of gyrotron system components location at ITER site

Power generating annex is the core of gyrotron system which is directly produced by Gycom. List of its deliverable components of includes at present 25 items. It does not include components of control system which stays a bit separately due to multiple issues associated mainly with permanent changes made by ITER organization (IO) in hardware and software. Test of the first two serial RF power sources as was agreed with IO were performed using prototype control system which essentially deviates from today's version. Its latest embodiment is to be manufactured this summer and tested in autumn together with fabric test of the 3rd serial gyrotron system.

Test results

At the factory acceptance test at Gycom two serial gyrotron as well as their analogue produced for F4E demonstrated very similar output parameters accumulated in following table

Gyrotron / Parameters	1st serial gyrotron	Gyrotron for F4E	2nd serial gyrotron
Frequency, GHz	169.9±03	170	169.9±03
RF output power, MW	0.97±0.05	0.92	0.96±0.05
Output radiation - HE_{11} mode content, %	97.3±0,5	96.8	Waiting for measurement
Efficiency (with depressed collector), %	56±3	56.3	53±3
Pulse duration, sec	1000	400/1000	1000
Beam voltage, kV	42.5 ± 0.4	42.5 ± 0.4	44.1± 0.4
Depression voltage, kV	27.5 ± 0.3	28.5 ± 0.3	26.5± 0.3
Beam current, A	42 ± 0.2	37.5 ± 0.2	42.4± 0.2
Body current, mA	10 ± 0.1	< 20	14 ± 0.1
Reliability, %	>95	> 80	> 95

Test program of serial gyrotrons included:

- 4 pulses 1MW/1000 s at duty cycle 0.25
- 10 pulses 1MW/75 s at duty cycle 0.05
- 40 pulses 1MW/100 s at duty cycle 0.25
- 20 pulses 1MW/500 s at duty cycle 0.25
- 3 pulses 1MW/200 s ON/OFF modulated /100 Hz
- 3 pulses 1MW/200 s ON/OFF modulated / 500 Hz
- 3 pulses 1MW/200 s ON/OFF modulated/1000 Hz

Pictures below illustrate recent test results of second serial gyrotron in more details. Fig. 2 shows signal traces monitored during one of 1MW/1000s pulse and graphs on the Fig.3 give statistics of power and frequency measurements for 500s and 1000s pulses.

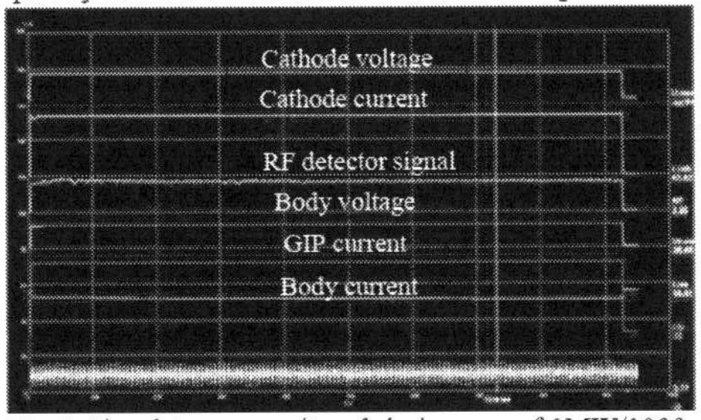

Fig. 2 Signal traces monitored during one of 1MW/1000s pulse

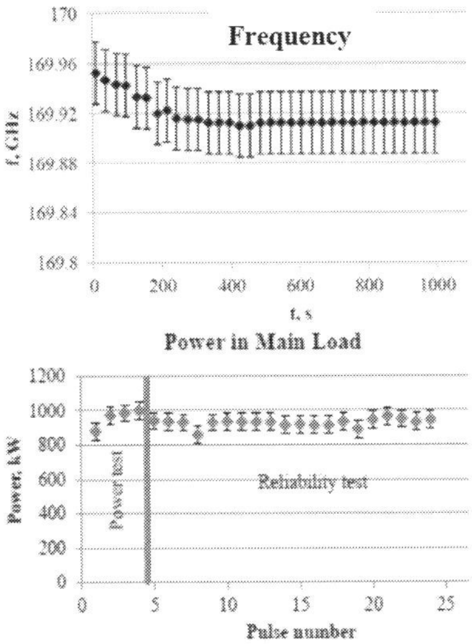

Fig.3 Statistics of frequency and power measurements for 500s and 1000s pulses

X-ray emission was found to be a serious issue even under testing of the first serial gyrotron. Lead shielding placed directly around gyrotron collector was not sufficient to suppress X-ray dose to admissible limit in vicinity of gyrotron casing due to stray radiation fraction going inside gyrotron body and penetrating through its copper wall. Problem was sold by mounting of rubber sheets with lead filling about outer contour of the support structure.

Gyrotron requested by F4E with ancillary components was delivered to SPC, mounted into SCM and connected to powering and cooling systems in June 2017. Under series of initial experiments in July-August 2017 at RF power of about 250kW inherent misalignment of terminal spherical load reflector with respect to waveguide one was revealed, which resulted in nonuniform load overheating. To correct misalignment Gycom in September 2017 produced and shipped to Lausanne copper bellows. In November 2017 combination of much higher RF power/ pulse duration was managed to attain (680kW/100 s and 880kW/10 s) but further increase of pulse energy was blocked by vacuum impairment in the load accompanied by light protection.

Test showed that produced to that time modifications of the load and its pumping port could not solve completely issue of load interior vacuum impairment. Additionally problem had arisen of correct power calorimetric measurements. Personal of F4E and SPC works now together to solve the problems. Test hopefully will be prolonged in July 2018.

Gyrotron system delivery issues

In the ITER site only case of building B15 has been erected but none interior infrastructure (cooling, powering, air conditioning etc.) has been set up. Site is not ready yet for start of any gyrotron system installation. "First plasma" operation is now postponed to 2027. IO is going to move delivery start of Russian gyrotron systems from 2018 to the end of 2019. To the moment of shipment implementation components of produced and passed the acceptance test gyrotron systems is to be stored in Gycom depositary being packed in the transport crates.

Summary

- Recently two gyrotrons systems for ITER ECH plant were manufactured and passed factory acceptance test successfully but being equipped with control system essentially deviating from the regular one
- Developing of regular control system for ITER RF power source is mainly delayed due to permanent changes in hardware and software initiated by IO and is waited now to be completed and tested in the end of this year
- Manufactured and tested gyrotron systems due to delay in ITER schedule are to be stored in Gycom depositary until at least end of 2019 when ITER site will hopefully be ready to accept package delivery and provide start of gyrotron system installation
- Analogue of 1MW/170GHz/1000s ITER gyrotron with agreed ancillary components was manufactured by Gycom under contract with F4E, passed factory acceptance test with the participation of F4E representatives and in June 2017 was shipped to Falcon facility of Swiss Plasma Center (SPC) in Lausanne
- Testing of Gycom gyrotron on site at SPC reveals problems with terminal load of the waveguide transmission line becoming evident by intensive outgassing accompanied by arc detection under increase of RF power or pulse duration Gyrotron test in SPC will be prolonged in July.

References

1. *L. Popov et al.* Super-High Power Gyrotrons for Electron-Cyclotron Plasma Heating // 10th International Workshop "Strong Microwaves and Terahertz Waves: Sources and Applications" N. Novgorod – Moscow, July 2017

2. *ITER_D_VJLBHF v1.1.* FAT report on 1st delivery RF DA RF power source

Status of ECRH experiments at GDT mirror trap

A. G. Shalashov[1,2], P. A. Bagryansky[2], E. D. Gospodchikov[1,2], L. V. Lubyako[1,2], Z. E. Konshin[2], V. V. Maximov[2], V. V. Prikhodko[2], V. Ya. Savkin[2], O. B. Smolyakova[1,2], A. L. Solomakhin[2], D. V. Yakovlev[2]

[1]Institute of Applied Physics RAS, Nizhny Novgorod, Russia, ags@appl.sci-nnov.ru
[2]Budker Institute of Nuclear Physics SB RAS, Novosibirsk, Russia

The paper summarizes the recent results of experiments on electron cyclotron resonance plasma heating (ECRH) and electron cyclotron emission (ECE) diagnostics at the GDT facility in Budker Institute, Novosibirsk. The design, physics of plasma confinement, and main goals of the machine are described in [1]. The gas-dynamic trap (GDT) is an axially-symmetric magnetic mirror device with a 7-m-long central cell and two expander cells at both ends (fig. 1). Previously we reported on plasma discharges with a very high temperature of bulk electrons achieved at this facility: up to 900 eV at the plasma density about 0.7×10^{19} m^{-3} [2]. This is more than a threefold increase with respect to previous experiments both at GDT and at other comparable machines. The breakthrough is made possible by the application of 0.7 MW 54.5 GHz ECRH in addition to standard 5 MW heating by means of neutral beam injection (NBI) [3-5].

While the previous experiments confirmed the core principles of longitudinal electron heat flux suppression in a gas-dynamic trap, it faced difficulties with mitigation of anomalous transport related to the development of MHD instability of the plasma column. In particular, when the microwave power was focused in a narrow axial plasma region thereby leading to a highly peaked electron temperature profile, the duration of effective heating was always limited to about 0.6 ms; later on, a flute instability developed preventing further absorption of microwaves.

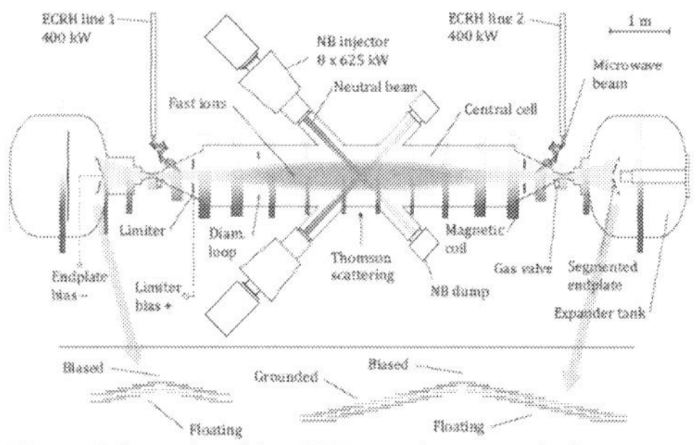

Fig. 1. Schematic of the GDT experiment. In the bottom we show new plasma-facing endplates for tuning of the radial distribution of plasma potential in the axial region.

During last experiments, we demonstrate the possibility of suppressing the MHD activity driven by localized ECRH using a new method for controlling the plasma differential rotation with the help of ring electrodes at the ends of the trap [6]. As a result, we demonstrate plasma discharges with parameters close to the reported record ones, but characterized by low MHD activity level and duration limited only by the available heating and magnetic confinement systems. In particular, a stable electron heating up to 450 eV during 1.5 ms period with ECRH power of 400 kW is attained for the first time at GDT.

A typical scenario is shown in fig. 2. The discharge is initiated in deuterium by electron cyclotron waves generated by second available gyrotron [7]. In the end of the start-up phase, the target plasma is heated with 5 MW NBI. At 7 ms, the additional 0.4 MW ECRH is switched on. Figure shows the evolution of on-axis electron temperature in the trap center for stabilized and non-stabilized discharges at otherwise identical experimental conditions. One see, when the end-plate potential U_{ep}<-400 V is applied, the electron temperature stays within the range of 350-450 eV for more than 1 ms while the plasma is MHD-stable throughout the whole ECRH duration, see "▲". Some decrease of the temperature at the end of the heating period is caused by detuning from optimal ECRH conditions due to the time variation of the magnetic field and drop in heating power at the end of the NBI pulse. Downward triangles "▼" illustrate cooling of bulk electrons immediately after ECRH is switched off while the stabilizing bias is still being applied. In agreement with previous observations, a discharge with on-axis ECRH but without the additional bias, ends in rapid degradation of plasma confinement, see "♦".

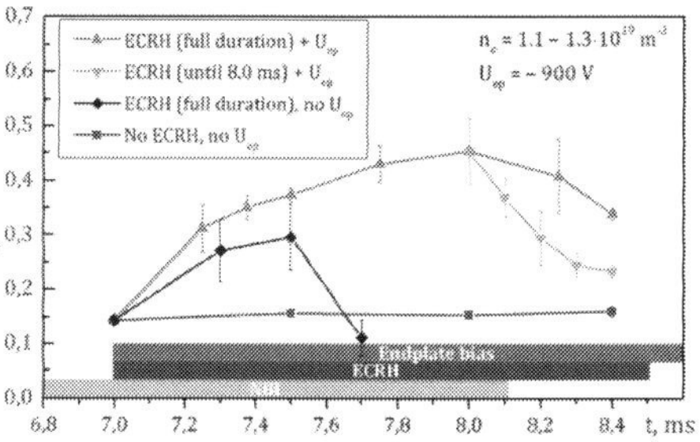

Fig. 2. On-axis electron temperature for: ▲stabilized full-time ECRH, ▼ stabilized ECRH with switch-off at 8 ms, ♦ not-stabilized ECRH, and ■ pure NBI discharges

Thus, we have shown experimentally that a local modification of the radial distribution of the plasma potential allows for a complete suppression of the MHD-instability caused by strong ECR heating and enables to maintain a high-T_e plasma in a large mirror trap. The minimum value of the external potential sufficient for stabilizing the plasma is quite modest (~400 V), so likely our technique may be extrapolated to higher temperatures and to larger devices without overwhelming engineering difficulties.

Attainment of quiescent discharges enables to perform more detailed studies of high-T_e transport physics not possible in MHD-unstable plasma. Of particular interest is the evolution of radial profiles of the electron temperature when ECRH is switched off, but the bias is maintained, see fig. 3 (top). In this case, the MHD effects in the core region of plasma can be excluded and the series of smoothly decaying profiles can be used to study plasma confinement [6]. An example for reconstructed radial profiles of electron energy fluxes, including ECRH "▲", electron heating from slowing-down of fast ions "▼" and gas-dynamic losses "■", recovered by fitting of transport equations to the experimental profiles is shown in fig. 3 (bottom). One finds that gas-dynamic losses are dominating, while the deposition of radial transport is less than 5% at each displayed point.

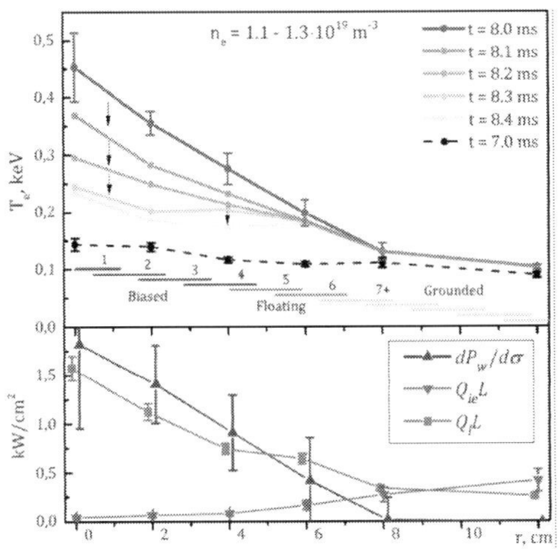

Fig. 3. Top panel: measured evolution of the electron temperature radial profiles at the trap center after ECRH switch-off. Bottom panel: reconstructed radial profiles of electron energy fluxes at $t = 8$ ms.

A new diagnostic system has been introduced at the GDT, allowing registration of plasma ECE at 75 GHz in a geometry corresponding to a possible scheme of ECR heating at the second harmonic of the extraordinary wave. With the help of this diagnostic, the spontaneous emission of fast electrons generated during microwave plasma heating is registered for the first time during the main phase of the plasma discharge, see fig. 4. According to the new data, the energy and lifetime of fast electrons are determined and the classical regime of their confinement is confirmed for the main phase of the plasma discharge. With the previous ECE system at the fundamental EC harmonic, these measurements were difficult since the emission of fast electrons was effectively screened by the bulk plasma [8].

Measured ECE data from thermal (bulk) electrons measurements is in reasonable agreement with theoretical expectations. However, to reproduce thermal ECE with a ray-tracing modeling, we must reduce the electron density as compared to the value measured independently at the central cross-section with TS diagnostics. This may indicate the influence of the ambipolar potential barrier formed by sloshing fast ions in the GDT conditions.

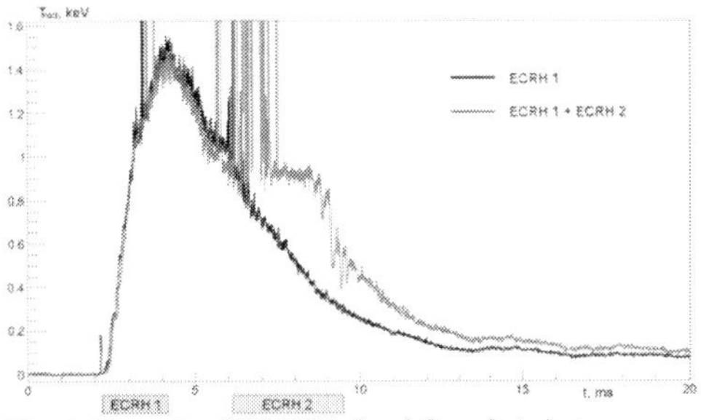

Fig. 4. Example of X2 ECE signal from fast electrons generated by one and two consecutive ECRH shots.

Next experimental campaign will be aimed at essential increase of fast ion confinement and neutron flux. New scenario is proposed theoretically with very flat ECRH power deposition profile that eventually provides high average T_e [9,10]. A new plug-in unit has been designed, manufactured and tested for GDT magnetic system upgrade for the planned experiment on ECR heating with the wide power deposition.

The work is supported by the Russian Scientific Foundation (pr. 14-12-01007).

References

1. *Ivanov A. A., Prikhodko V. V.* Gas-dynamic trap: an overview of the concept and experimental results // Plasma Phys. Control. Fusion. 2013. V. 55. P. 063001.

2. *Bagryansky P. A., Shalashov A. G. et al.* Threefold increase of the bulk electron temperature of plasma discharges in a magnetic mirror device // Phys. Rev. Lett. 2015. V. 114. P. 205001. (arXiv:1411.6288).

3. *Bagryansky P. A., Kovalenko Yu. V., Savkin V. Ya. et al.* First results of an auxiliary electron cyclotron resonance heating experiment in the GDT magnetic mirror // Nucl. Fusion. 2014. V. 54. P. 082001.

4. *Bagryansky P. A., Gospodchikov E. D., Kovalenko Yu. V. et al.* ECR heating experiment in the GDT magnetic mirror: recent experiments and future plans // Fusion Sci. Technol. 2015. V. 68. P. 87.

5. *Bagryansky P. A, Anikeev A. V., Denisov G. G. et al.* Overview of ECR plasma heating experiment in the GDT magnetic mirror // Nucl. Fusion. 2015. V. 55. P. 053009.

6. *Yakovlev D. V., Shalashov A. G., Gospodchikov E. D. et al.* Stable confinement of high-electron-temperature plasmas in the GDT experiment // Nucl. Fusion. 2018. (https://doi.org/10.1088/1741-4326/aacb88)

7. *Yakovlev D. V., Shalashov A. G. et al.* Electron cyclotron plasma startup in the GDT experiment // Nucl. Fusion. 2016. V. 57. P. 016033. (arXiv:1607.01051).

8. *Shalashov A. G., Solomakhin A. L. et al.* // Phys. Plasmas. 2017. V. 24. P. 082506. (arXiv:1611.09137).

9. *Shalashov A. G., Balakin A. A. et al.* Quasi-optical theory of microwave plasma heating in open magnetic trap // Phys. Plasmas. 2016. V. 23. P. 112504.

10. *Shalashov A. G., Balakin A. A., Khusainov T. A. et al.* Quasi-optical simulation of the electron cyclotron plasma heating in a mirror magnetic trap // JETP. 2017. V. 124, No 2, P 325.

New developments in the field of high current ECR ion sources at the IAP RAS

V.A. Skalyga, S.V. Golubev, I.V. Izotov, M.Yu. Kazakov, R.L. Lapin, S.V. Razin, A.V. Sidorov, R.A. Shaposhnikov, A.F. Bokhanov

Institute of Applied Physics of Russian academy of Sciences, Nizhny Novgorod, Russia, skalyga@ipfran.ru

Investigations of pulsed ECR discharge in an open magnetic trap under conditions of powerful ECR heating with gyrotron mm-waveband radiation were carried out over the last 20 years at the Institute of Applied Physics (IAP RAS, Nizhniy Novgorod, Russia) [1-5]. In the beginning the work was devoted to development of a high frequency ECR source of multi-charged ions with outstanding parameters of plasma heating (37.5 GHz, 100 kW). According to Geller's scaling laws [6] such increase in frequency and power in comparison to conventional ECRIS was expected to boost the ion source performance and provide a significant progress in ECRIS development. However, due to short pulse operation mode and low repetition rate of the used gyrotrons (pulse duration < 1 ms, 0.1 Hz) breakdown and discharge conditions similar to a conventional ECRIS were unreachable. The minimum neutral gas pressure was two orders higher (10^{-4} mbar) and the plasma parameters differed significantly from conventional ECRIS. After years, this work resulted in development of a new type of ion source – high current gasdynamic ion source.

Experimental studies of gasdynamic ECRIS were performed at SMIS 37 facility, schematically depicted in Fig. 1. The plasma was created by 37.5 and 75 GHz gyrotron radiation with power up to 100 kW. Gasdynamic ECRIS has demonstrated a good performance producing high current (100-300 mA) multi-charged ion beams with moderate average charge (Z=4-5 for argon). Gasdynamic ECRIS has appeared to be especially effective in low emittance hydrogen and deuterium beams formation. Proton beams with current up to 500 emA, RMS emittance below 0.07 π·mm·mrad have been demonstrated in recent experiments.

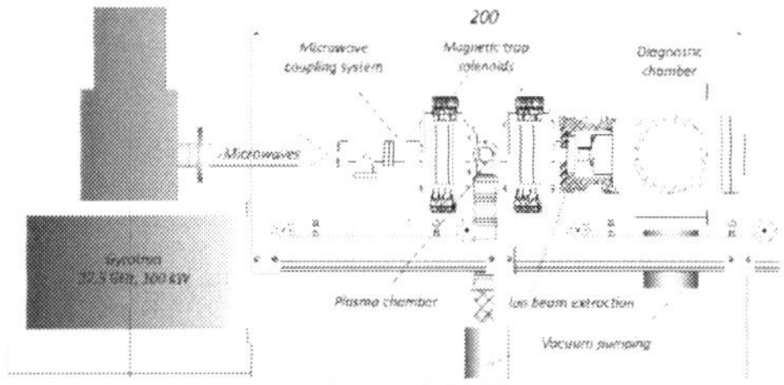

Fig. 1. SMIS 37 experimental facility.

Experiments aimed to demonstrate the benefits of the high current gasdynamic ECR ion source as a part of D-D neutron generator were

Measurement of the produced neutron flux was performed by bombarding a heavy ice target with 300 emA D^+ beam accelerated to 45 keV energy. The D_2O target yield was 10^9 s^{-1}. It is worth noting that the area of the target was about 1 cm^2, which corresponds to a neutron flux density of about 10^9 $cm^{-2}\cdot s^{-1}$ at the target.

As it was demonstrated a high-current ECR ion source with quasi-gasdynamic plasma confinement and heating with gyrotron microwave radiation allows the formation of light ion beams with uniquely low emittance (for a given current level). Such a low emittance enables focusing of the ion beam into a small spot. So it could be used as alternative approach for creation of a point-like neutron source for neutronography. A magnetic lens was used for beam focusing as the next step. Measurements have shown that lens focus for 1.3 T field was situated approximately in 10 cm from lens center. Example of scintillator luminescence image induced by 80 keV, 60 mA deuterium beam is presented in fig. 2. The best result for focused beam full width at half maximum was 0.75 mm.

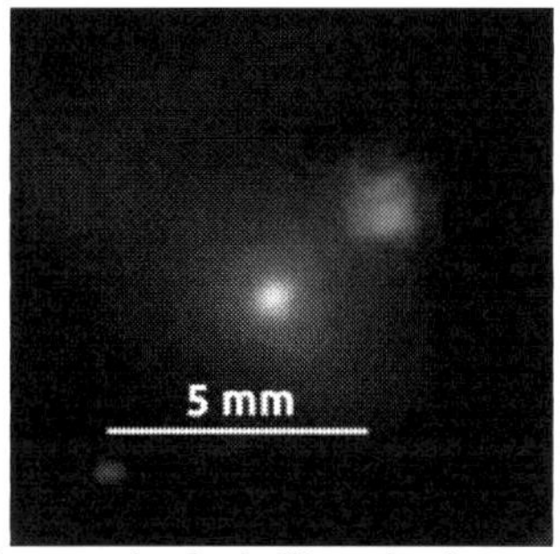

Fig. 2. Photograph of scintillator luminescence (green glow) induced by 80 mA hydrogen beam in magnetic lens focal plane (red glow corresponds to optical emission from plasma through the plasma electrode aperture)

Obtained focused ion beam was used for TiD_2 target bombardment. As a result the "point-like" neutron emission with total yield about 10^{10} s^{-1} have been demonstrated.

Another direction for experimental studies is H^- ion source development. H^- ion sources are of great demand for charge exchange injection into cyclotrons and storage rings. It has been recently demonstrated that a gasdynamic ECR ion source based on ECR discharge in a simple mirror trap is very efficient for proton beam production [7,8]. Therefore, it was suggested to use the gasdynamic plasma source as the first stage of an H^- ion source based on volumetric production through dissociative electron attachment [9,10] (DEA).

The first experiments were performed in a pulsed mode with 37 GHz / up to 100 kW gyrotron radiation in a dual-trap magnetic system, consisting of two identical simple mirror traps. The schematic of the experimental facility is presented in fig.3. The first trap was used for plasma production under ECR conditions. Dense hydrogen plasma flux (with estimated current density of about 1-5 A/cm^2) from the first trap was allowed to flow into the second trap through a perforated conducting plate. The grid placed between the traps prevents microwave leakage into the second part to avoid electron heating there. Such configuration was chosen to produce a flux of "hot" electrons (energy 50 – 100 eV), which can effectively ionize and excite high vibrational states of hydrogen molecules in the second trap, and to produce anions there as the result of DEA with "cold" electrons (energy < 10 eV). Thus, the plasma confined in the second trap presumably consists of two electron populations enabling the volume production of H^- ions. Such approach is similar to the one suggested in [11] where heating 2.45 GHz ECR discharge was used as a plasma cathode producing "hot" electrons.

In case of 1 mm diameter of the hole for beam extraction negative ion current was up to 0.7 mA. The corresponding ion current density was 80 mA/cm^2. Such results makes further investigations very perspective.

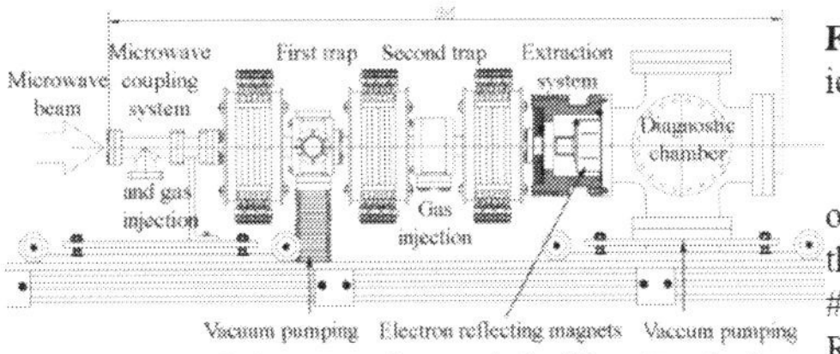

Fig. 3. Scheme of the experimental facility. Typical negative ion current waveform. Microwave power 80 kW, constant gas flow into the first trap and pulsed injection into the second one.

The main part of previous experiments were carried out in a pulsed operation mode. Preliminary studies of plasma parameters were performed using a CW source with 24 GHz/5 kW gyrotron heating. Obtained experimental results have demonstrated that all gasdynamic source advantages could be realized in CW operation. To continue development of a CW gasdynamic ion source a new experimental facility named GISMO (Gasdynamic Ion Source for Multipurpose Operation) is under construction at the IAP RAS. Future facility have been named GISMO (Gasdynamic Ion Source for Multipurpose Operation). This facility is aimed to produce continuous high-current (>200 mA) ion beams with low emittance (<0.2 π·mm·mrad). The scheme of the future experimental facility is show in figure 1. The key elements of the setup are 28 GHz/10 kW and 37,5 GHz/20 kW CW gyrotrons manufactured by Gycom [12]. This microwave generators will be equipped with power supplies suitable for CW or pulsed operation. A fully permanent magnet magnetic trap will be used for plasma confinement. Magnetic field configuration was designed to be similar to a simple mirror trap close to the system axis with field strength at magnetic mirrors of 1.5 T and mirror ratio close to 6. Distance between magnetic mirrors is about 12 cm. For ion beam extraction it is planed to use 3 or 4-electrode system with maximum acceleration voltage up to 100 kV. Such extraction requires development of an appropriate high-voltage insulation of the discharge chamber from other parts. In this regard, one of the key elements of the installation is the DC-break of the microwave transmission line. It was proposed to be implement a quasioptical system shown in Fig. 1. Plasma chamber will be 30 cm in length and 4 cm in diameter. It is equipped with water cooling along whole surface from the coupling system to the flange.

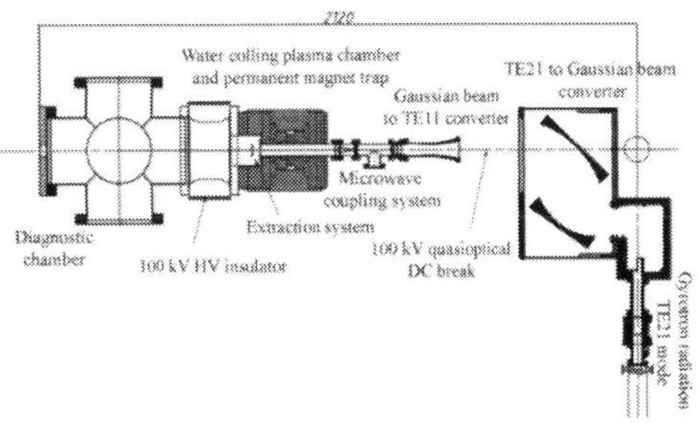

Fig. 4. Scheme of the GISMO 28/37 CW high current ion source.

The work is being realized in frames of realization of Federal targeted program R&D in Priority Fields of the S&T Complex of Russia (2014-2020) contract #14.604.21.0195 (unique identification number RFMEFI60417X0195).

References

1. *S.V. Golubev et al.*, Rev Sci. Instrum. 75, 1675 (2004).
2. *S.V. Golubev et al.*, Trans. Fus. Sci. and Tech. 47, 345 (2005).
3. *A. Sidorov et al.*, Rev Sci. Instrum. 77, 03A341-1 (2006).
4. *V. Skalyga et al.*, Plasma Sources Sci. Tech. 15, 727 (2006).
5. *S. Golubev et al.*, Nucl. Instrum. Meth. in Phys. Research B 256, 537 (2007).
6. *R. Geller*, Electron cyclotron resonance ion sources and ECR plasmas (Institute of Physics, Bristol, 1996).
7. *V.A. Skalyga et al.*, Rev. Sci. Instrum. **85**, 02A702-1-3 (2014).
8. *V. Skalyga et al.*, Rev. Sci. Instrum. **87**, 02A716 (2016)
9. *M. Bacal, G.W. Hamilton*, Phys. Rev. Lett. **42**, 1538 (1979).
10. *M. Bacal* Nuclear Fusion **79**, 02A516 (2008).
11. *R. Keller et al.*, in AIP Conf. Proceedings **773** (American Inst. Of Physics, Melville, NY. 2005), p. 104–106.
12. www.gycom.ru

2018 Status of the Measurement Capabilities for Fusion Gyrotrons at KIT/IHM

T. Ruess[1], K. A. Avramidis[1], M. Fuchs[1], G. Gantenbein[1], S. Illy[1], Z. Ioannidis[1], F. C. Lutz[1], S. Ruess[1,2], T. Rzesnicki[1], M. Thumm[1,2], D. Wagner[3], J. Weggen[1] and J. Jelonnek[1,2]

[1]IHM, [2]IHM Karlsruhe Institute of Technology (KIT), Kaiserstr. 12, D-76131 Karlsruhe, Germany, tobias.ruess@kit.edu
[3]Max Planck Institute for Plasma Physics, Boltzmannstr. 2, D-85748 Garching, Germany

Abstract - **This paper gives an overview about three key measurement capabilities at KIT/IHM for fusion gyrotrons. The discussed topics are: RF window test facility, low power launcher test setup and the frequency measurement setup placed at the high power test facility. Further, approaches for upgrading the setups to frequencies above 200 GHz are discussed.**

Introduction

Measurements are needed to evaluate the operation of single gyrotron components itself and further, the assembled gyrotron. This paper summarizes three main measurement facilities for fusion gyrotrons at the KIT/IHM in the year 2018 and possible approaches for upgrading these facilities towards operation above 200 GHz are discussed.

Firstly, the RF window measurement test facility is described which gives the possibility to characterize the RF output window in terms of transmission and reflection features. Secondly, the quasi-optical system consisting of launcher and mirrors can be verified after the fabrication in the low power measurement test facility. Thirdly, the frequency measurement setup is a key instrument for the validation of the behavior of the gyrotron in the high power measurement facility.

Window Measurement Setup

The RF-window requirements are high temperature conductivity and high RF-power transmission capacity. Chemical vapor deposition (CVD) diamond shows the best properties for this application, so far. The major properties of CVD diamond are: relative permittivity ε_r = 5.67, loss tangent $\tan\delta \approx 2\cdot10^{-5}$ and thermal conductivity $\kappa \approx$ 1900 W/m·K. The window measurement setup at the KIT/IHM is presented in [1]. The configuration of the setup is a four port scenario. One port is used for transmitting and one for receiving the signal (reflectivity measurement). A beam splitter is required to separate the transmitting (Tx) and receiving (Rx) path. The two unused ports have to be matched perfectly. The matching is realized using absorber material and a ceramic plate which is tilted by the corresponding Brewster angle. The dynamic range of the system is calculated to 60 dB and the covered frequency range is from 110-170 GHz and is determined by the used frequency multiplier (Tx path) and harmonic mixer (Rx path). The setup can be easily adapted to higher frequencies by changing the above mentioned multiplier and harmonic mixer.

Launcher Measurement Setup

This setup is used to verify the quasi-optical (q. o.) gyrotron components (launcher and mirrors) at low power (~1 mW). The availability of such a test system is vital for the gyrotron design because the q. o. output coupler is one of the most critical gyrotron component. This q. o. system converts the rotating cavity mode into a linearly polarized Gaussian beam. Compared with the gyrotron cavity, the low-power test setup is operating without electron beam. Therefore, the correct mode has to be excited in a different way using a mode generator, as depicted in Fig. 1. The design is based on a perforated gyrotron cavity as published in [2].

The RF wave is transmitted using a horn antenna. A satisfying mode is only excited if the wave beam has a minor phase changing. For this reason, a lens system is required to convert the spherical wave into a Gaussian like beam. The quasi-parabolic mirror is necessary to focus the wave on the caustic radius of the preferred mode in the cavity. The cavity is specially designed for this purpose using a scattering matrix code [3]. As in coaxial gyrotrons, an insert is used for mode selection and easier excitation of the mode [4,5]. The midsection of the cavity wall is perforated so that the wave can propagate through the holes.

The excitation of the correct mode is the most challenging task due to the high amount of degrees of freedom of the system. Therefore, the excited mode has to be checked first. For this reason, a non-linear up taper is mounted at the end of the cavity for reaching a better measurement resolution due to the increased diameter. If the mode is excited with sufficient purity, the non-linear up taper will be replaced by the launcher. The excited mode (cavity with non-linear up taper or Gaussian beam using the quasi-optical system) is measured with a nearfield pick-up antenna. The antenna is mounted on a 3D measurement arm, where a stepwise scan of the complete plane can be done.

The KIT has developed, in close collaboration with the IPP Garching, mode generators for several operating gyrotron modes ($TE_{22,6}$-, $TE_{22,8}$-, $TE_{28,8}$-, $TE_{32,9}$-, $TE_{34,19}$-mode). Recently, an upgrade of the 1 MW hollow-cavity $TE_{28,8}$-mode gyrotrons operating at Wendelstein 7-X (W7-X) towards 1.5 MW RF output power is under consideration ($TE_{28,10}$ mode) [6].

Developments are ongoing targeting more precise measurements and to implement an intelligent algorithm achieving a substantial time saving for finding the correct mode in high quality. Further, the RF front end is upgraded to cover the frequency range from 140 – 330 GHz for future gyrotron components.

Frequency Measurement Setup

Parts of the gyrotron signal is received by a pick-up antenna at the tube's relief window. Then, the signal is divided and feeds two different frequency measurement systems. One of them is called FMS (frequency measurement setup) [7] and the other one is called PMS (pulse spectrum analysis system) [8].

At the beginning of the FMS, the signal is down converted by a harmonic mixer. The mixer operates on the 9th harmonic and has an intermediate frequency (IF) range of f_{IF} = 0.1-18 GHz. The frequency of the IF signal is determined using a filter-bank spectrometer and a frequency time analyzer. The filterbank spectrometer is constructed with multiplexers which divide the 18 GHz spectrum into 9 sub-channels of ~ 2 GHz bandwidth. Afterwards the signals are detected by wide-band RF diodes and sampled with an oscilloscope. The dynamic range is 24 dB. The filterbank spectrometer gives a broadband but not precise overview about the gyrotron behavior. Therefore, a modulation domain analyzer is implemented in this channel achieving high frequency resolution in a limited frequency bandwidth of around 2 GHz. This path offers a dynamic range of 20 dB. The FMS can be easily upgraded by changing the first mixer stage to cover any frequency ranges (e.g. 170-260 GHz).

The FMS is a very useful setup to get a very quick overview about the behavior of the gyrotron operation. The major drawbacks of the FMS setup are the low dynamic range and the limited bandwidth, where measurements can be performed with high resolution. Therefore, the PSA has been implemented [8]. This system has the following aims: (i) high dynamic range for parasitic oscillation detection, (ii) high frequency resolution for the diagnosis of frequency instabilities and modulation, (iii) high bandwidth for mode competition investigations (mode switching) and (iv) time dependent analysis.

The measurement setup is based on a super-heterodyne receiver. Ambiguities can occur in the RF spectrum reconstruction due to the lack of knowledge if the RF signals are placed in the upper or lower side bands. For this reason, a second identical channel with slightly different LO (local oscillator) frequency is designed. The effective bandwidth f_{BW} of the measurement setup can be increased from f_{BW} = 3-3.5 GHz to $f_{BW} \approx 6$ GHz using the knowledge about the evaluation of both side bands. The effective bandwidth is further increased to $f_{BW} \approx 12$ GHz using a second identical channel with two more mixers. The dynamic range of this setup amounts to 60 dB, which is enough to detect eventual parasitic oscillations.

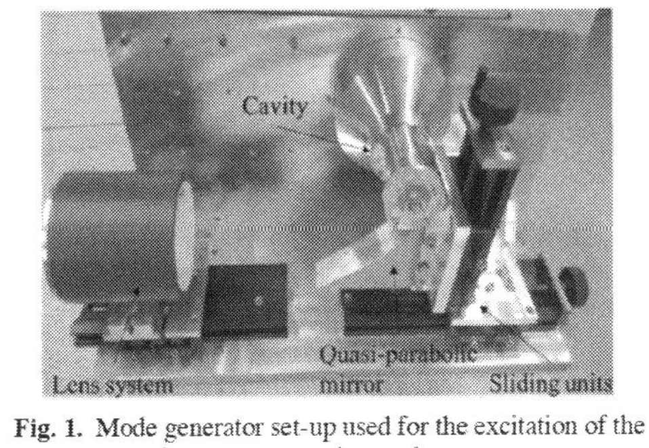

Fig. 1. Mode generator set-up used for the excitation of the correct operating gyrotron cavity mode.

A possible upgrade is by shunting a second mixer, which operates in a frequency range of from 170-260 GHz, as shown in Fig. 2. A diplexer and attenuators can be used to separate the different frequency ranges. Shunting several mixers is a first proposal for upgrading the existing PSA for DEMO relevant gyrotrons operating at frequencies up to 240 GHz.

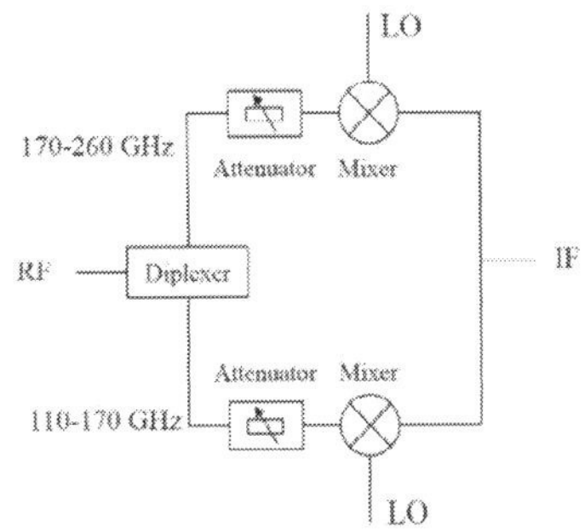

Fig. 2. Possible upgrade of the PSA towards a frequency range from 110-260 GHz.

References

[1] *J. Flamm, et al.*, "Characterization of windows for fusion applications using a D-band network analyzer", *33rd Int. Conf. on Infrared, Millimeter and Terahertz Waves*, Pasadena, CA, 2008, pp. 1-2.

[2] *N. L. Alexandrov, et al.*, "Selective excitation of high-order modes in circular waveguides", *Int. J. Infrared and Millimeter Waves*, **13**, 1369-1385, 1992.

[3] *D. Wagner, et al.*, "Analysis of a complete gyrotron oscillator using the scattering matrix description", *Int. J. Infrared and Millimeter Waves*, **19**, 185-194 (1998).

[4] *C. T. Iatrou, S. Kern, A. B. Pavelyev*, "Coaxial cavities with corrugated inner conductor for gyrotrons", *IEEE Trans. Microwave Theory Tech.*, **44**, pp. 56-64, 1996.

[5] *K. A. Avramides, et al.*, "Design Considerations for powerful continuous-wave second-cyclotron-harmonic coaxial-cavity gyrotrons", *IEEE Trans. on Plasma Science*, **32**, no. 3, 2004.

[6] *K. A. Avramides, et al.*, "Studies towards an upgraded 1.5 MW gyrotron for W7-X", *20th Joint Workshop on Electron Cyclotron Emission (ECE) and Electron Cyclotron Resonance Heating (ECRH)*, May 2018, Greifswald, Germany.

[7] *T. Geist, M. Thumm, W. Wiesbeck*, "Contiguous filterbank receiver for a pulsed 140 GHz gyrotron, *Conf. Digest, 16th Int. Conf. on Infrared and Millimeter Waves*, Lausanne, 1991, Proc., SPIE 1567, pp. 274-275.

[8] *A. Schlaich, et al.*, "Transient millimeter-wave signal analysis with unambiguous RF spectrum reconstruction", *IEEE Trans. on Microwave Theory and Techniques*, vol. 61, no. 12, pp. 4660-4666, Dec. 2013.

Thirty Joint Russian-German Meetings on ECRH and Gyrotrons - A Retrospect -

Manfred Thumm[1,2],

[1]Institute for Pulsed Power and Microwave Technology (IHM),
[2]Institute of Radio Frequency Engineering and Electronics (IHE),
Karlsruhe Institute of Technology (KIT), 76131 Karlsruhe, Germany, manfred.thumm@kit.edu

Foundation of Russian-German Cooperation on ECRH and Gyrotrons

The USSR-FRG scientific-technical cooperation in the field of electron cyclotron resonance heating (ECRH) and gyrotrons was founded in 1988, when delegations of the German Ministry for Research and Technology (BMFT) (Dr. S. von Krosik) and of the USSR State Committee for Utilization of Atomic Energy (Dr. M.P. Moiseeva) met at the I.V. Kurchatov Institute of Atomic Energy (KIAE) in Moscow to establish a cooperation in the area of thermonuclear fusion. Leading scientific members of the German delegation were Prof. R. Wienecke (Institute for Plasma Research (IPF), University of Stuttgart) and Dr. G. Grieger (Max Planck Institute for Plasma Physics (IPP), Garching). The USSR delegation was represented by Dr. B.B. Kadomtsev, Dr. G.A. Eliseev and Dr. W.S. Strelkov (KIAE) and Dr. V.A. Flyagin (Institute of Applied Physics (IAP) of the Soviet Academy of Sciences, Gorky). In a letter of Dr. Flyagin to Prof. Wienecke dated June 6, 1988, V.A. Flyagin gave the information, that "in the future we shall be able to organize seminars in the USSR".

The First Meetings in 1989 and 1990

The topics of the first USSR-FRG Meeting on ECRH and Gyrotrons were discussed during a scientific stay of Dr. G.S. Nusinovich at Kernforschungszentrum Karlsruhe (KfK) in July 1988. The preliminary agenda of the workshop was:

-Status reports and future plans on ECRH system development and on ECRH experiments
-Special subjects of gyrotron development
  -Design of Magnetron Injection Guns (MIGs)
  -Whispering gallery vs. circular mode gyrotrons
  -Dielectric vacuum windows, material and design
  -Gyrotron operation experience
  -Alternative millimeter (mm)-wave sources
-Advances in electrodynamic systems
-Development perspectives of transmission lines
  -Quasi-optical (q.o.) vs. oversized waveguides.

The first meeting took place from March 6-10, 1989 in Germany at KfK Karlsruhe, IPF Stuttgart and IPP Garching. The USSR delegation had 5 members: G.G. Denisov, V.A. Flyagin, G.S. Nusinovich, M.I. Petelin from IAP Gorky and A.A. Borschegovsky from KIAE Moscow. The approximately 20 German participants came from KfK Karlsruhe, IPF Stuttgart, IPP Garching, RWTH Aachen and Technical University of Hamburg-Harburg. The meeting was accompanied by laboratory visits at KfK, IPF and IPP.

Since 1989 this series of annual meetings is alternatively organized in Germany and Russia. The second meeting was held at KIAE Moscow and IAP Gorky. The participating institutions were: KIAE Moscow, Institute of General Physics Moscow, IAP Gorky, IPP Garching, IPF Stuttgart and KfK Karlsruhe. The FRG delegation consisted of five members: V. Erckmann from IPP, W. Kasparek and M. Thumm from IPF and G. Dammertz and H.-U. Nickel from KfK. The total number of participants was about 50. 22 reports were presented and discussed. Technology items were: long-pulse 140 GHz gyrotrons, cryogenically cooled windows, combined q.o. and waveguide transmission lines. In the field of plasma physics, T-10 and W7-AS ECRH and electron cyclotron current drive (ECCD) experiments were discussed. Visits were organized to the tokamaks T-10 and T-15 at KIAE and to the laboratories involved in the development of gyrotrons, relativistic microwave devices, electrodynamic systems and plasma physics at IAP. In 1991, the city of Gorky was re-named as Nizhny Novgorod.

Agreement on Cooperation between IAP and IPP

During the 3rd meeting of this workshop series from June 12-18, 1991 at Karlsruhe, Stuttgart and Garching in Germany, the Russian delegation reported about gyrotrons, CARMs, electrodynamic systems, and ECRH&CD on T-10, whereas the German team presented results on transmission lines, windows, electrodynamic systems, and ECRH&CD on the stellarator W7-AS. On course of this meeting, on June 18, 1991, an Agreement of Cooperation between IAP and IPP was signed aiming at the use of 0.5 MW, 140 GHz Soviet gyrotrons with 0.5 s pulse duration at W7-AS and ASDEX Upgrade. The two parties of the Agreement decided that the first stage of the cooperation should cover in particular the following fields:

-Test of Soviet high power gyrotrons at the W7-AS stellarator with respect to frequency stability, mode purity and power control (in collaboration with KfK and KIAE);
- Tests of the q.o. transmission line developed by IPF with respect to power capability, losses and transmission efficiency (in collaboration with IPF);
-Tests of microwave barrier windows under high power conditions;
-Plasma physics experiments (in collaboration with KIAE, IPF and KfK): plasma generation and 2nd harmonic X-mode heating in W7-AS at 140 GHz, ECCD, on-and off-axis heat-wave experiments and combined ECRH and NBI.

Later, this cooperation led to the delivery of commercial 140 and 140/105 GHz GYCOM gyrotrons to W7-AS and ASDEX Upgrade.

Scientific and Technological Highlights of the Thirty Joint Russian-German Meetings

A series of novel developments have been achieved during this Russian-German cooperation.

Plasma Physics

ECRH&CD experiments have been performed in the tokamaks T-10 and ASDEX-Upgrade, in the stellarators W7-AS and W7-X, and in the gas dynamic mirror trap (GDT), leading to substantial progress in understanding the physics of fusion plasmas.

T-10: 1st and 2nd harmonic ECRH&CD experiments with power levels up to 3 MW; ECCD with reversed magnetic shear; demonstration of internal transport barrier formation by high-efficiency off-axis ECRH, ECRH&CD experiments on density pump-out; heavy-ion impurity control and influence of plasma reflections on gyrotron operation.

ASDEX-Upgrade: 2nd harmonic ECRH&CD experiments; complete stabilization of (3/2) and (2/1) neoclassical tearing modes (NTMs) by ECCD, record temperatures (T_e, T_i > 13 keV) with ECCD sustained thermal transport barrier; electron heat transport (steady state and transient phase) explained by ion-temperature-gradient (ITG) / trapped-electron-mode (TEM) turbulence after stabilization of (3/2)-NTM at β_N = 3.1; consistent experimental and theoretical explanation of flattened density profiles (pump out) under intense ECRH; first operation of industrial 1 MW, 10 s two-frequency gyrotrons; H-mode discharges with ITER-relevant safety factor using 3rd harmonic ECRH; 4.4 MW ECRH&CD using 8 gyrotrons for ELM mitigation and NTM stabilization; successful operation of a compact fast-directional switch (FADIS) in the ECRH transmission system; advanced tokamak experiments in W-coated ASDEX-Upgrade and non-inductive H-mode operation with safety factor-(95) = 5.3; excellent performance of new two-frequency gyrotrons for the ECRH-3 system.

W7-AS: Discovery of H-mode discharges using 70 GHz 1st harmonic O-mode and 140 GHz 2nd harmonic X-mode ECRH; installation of Collective Thomson Scattering (CTS) diagnostics and first results on CTS ion temperature measurements; investigation of O2-mode heating and mode conversion schemes; first experiments on O-X-B conversion heating, plasma heating and highly efficient current drive via Bernstein waves in over-dense plasmas (> $3x10^{20}/m^3$), achievement of 7 keV electron temperature with 2.7 MW ECRH; introduction of electron Bernstein wave emission diagnostics; extreme power densities in ECCD experiments and development of related non-linear theoretical description.

W7-X: Plasmas generated and heated by X2-mode ECRH only; parallel operation of ten 1 MW, 140 GHz gyrotrons and a unique purely q.o. transmission line; successful demonstration of O2-mode operation above the X2-cutoff at plasma densities up to $1.4x10^{20}/m^3$, in combination with pellet injection a stellarator record triple product of $0.66x10^{20}$ m^{-3} s keV and T_e = T_i = 3.9 keV were achieved; successful first operation of two remote steering launchers at; commissioning and first results of the CTS diagnostic.

GDT: 0.8MW, 54GHz ECH doubles the neutron rate.

Gyrotrons and ECRH&CD Technology

Gyrotron, transmission line and ECRH&CD technology development has been done at IAP, IPF (now IGVP), IPP and KfK (later FZK and now KIT).

Gyrotrons: Introduction of single-stage depressed gyrotron collector (SDC) and additional 50 Hz transverse electron-beam sweeping; demonstration of 70 % efficiency and 99 % output-mode purity of an industrial 75 GHz gyrotron; 0.2 MW, 84 GHz, continuous wave (CW) gyrotron for the LHD stellarator at NIFS; megawatt-class 110 GHz gyrotron for TEXTOR (pulse length 2 s), 140 GHz gyrotrons for T-10, ASDEX-Upgrade (10 s), W7-AS (3 s) and W7-X (1800 s) and 170 GHz gyrotrons for ITER (1000 s, 1 kHz power modulation); 1.2 MW, 100 s, 52 % efficiency operation of the Russian ITER tube and corresponding 1.5-2 MW development for ITER Upgrade; development of short-pulse, coaxial-cavity gyrotrons: 1.5 MW, 140 GHz with dual-beam output coupler and two windows, 2.2 MW, 165 GHz with 49 % efficiency at 2 MW (with SDC), 2.3 MW, 170 GHz with 46 % efficiency at 2.2 MW (with SDC); first short-pulse 114-162 GHz, 1 MW gyrotrons with step-frequency tuning using Brewster output windows and experiments on fast step-tuning using a hybrid gyrotron magnet; 1 MW two-frequency and multi-frequency gyrotrons; future DEMO gyrotrons and the development of other types of mm- and sub-mm-wave sources for non-fusion applications were discussed.

ECRH&CD Technology: Introduction, test and application of Chemical Vapor Deposition (CVD) diamond and doped silicon as high-power window materials; due to the superior mechanical characteristics, synthetic diamond is the final choice; development, test and implementation of highly efficient broadband quasi-optical mode converters, diplexers (FADIS), polarizers, directional couplers; FADIS switching of 0.5 MW at 25 kHz; successful 0.9 MW, 30 min. demonstration of the quasi-optical multi-beam transmission line of W7-X; development and successful high-power test of the remote-steering launchers at W7-X; fast synthesis of wideband multi-mode waveguide components, development, test and implementation of corrugated, high-mode purity HE_{11}-hybrid mode transmission line components and dummy loads; integration of the ITER upper launcher.

Non-Fusion Applications of Gyrotrons

Application of highly-efficient gyrotrons in ECR heavy ion sources and in technological installations for microwave materials processing as sintering of advanced ceramics and metal-powder compacts as well as for microwave-assisted fabrication of low-loss, large-size CVD diamond disks. Studies on non-thermal microwave effects.

Development and application of frequency-stable THz gyrotrons in Dynamic Nuclear Polarization (DNP) spectroscopy Nuclear Magnetic Resonance (NMR) experiments, for plasma diagnostics and for detection of concealed radioactive materials.

Development of pulsed broadband gyro-TWTs and CW gyro-BWOs with helical interaction circuit for RADAR applications and materials processing.

PIC Simulations of Ka-band Ultra-Short Pulse Oscillator with Resonance Cyclotron Absorber in the Feedback Loop

N.S. Ginzburg[1,2], G.G. Denisov[1], M.N. Vilkov[1], I.V. Zotova[1], A.S. Sergeev[1], S.V. Samsonov[1], A. Marek[3], and J. Jelonnek[3]

[1]Institute of Applied Physics RAS, Nizhny Novgorod, Russia, vilkovmn@ipfran.ru,
[2]Nizhny Novgorod State University Nizhny Novgorod, Russia
[3]Karlsruhe Institute of Technology, Karlsruhe, Germany

In [1,2] it was shown that periodical trains of ultrashort microwave pulses (USP) can be produced in a two-section microwave oscillator consisting of an amplifier and a nonlinear (saturable) absorber in the feedback loop. In such a system, the process of pulse formation is based on the effect of passive mode-locking which is well known in laser physics [3]. At the present moment, the Ka-band USP oscillator with a broadband helical-waveguide gyro-TWT [4] as an active unit is under development at the Institute of Applied Physics. The saturable absorber will be realized in an auxiliary section based on the cyclotron resonance interaction with an initially rectilinear magnetized electron beam. In this case the nonlinearity of absorption is provided due to relativistic dependence of a gyrofrequency on the electrons energy. The basic parameters of the USP oscillator have been defined on the basis of a simple averaged model. This paper is devoted to the full-scale particle-in-cell (PIC) simulations of such a system based on the software CST Studio Suite [5] aimed to optimize the parameters taking into account the real conditions of the planning experiment.

For simulations the computer model has been developed, in which the helical-waveguide amplifier and the cyclotron absorber were installed opposite each other (see Fig. 1). The gyro-TWT operates at the second cyclotron harmonic while the fundamental cyclotron resonance is used in the absorber unit. From the cathode ends both sections were bounded by resonance reflectors. The independent electron beams with the given parameters were entered the calculated space from the left and right ends of the corresponding sections and, after interaction, deflected in a strong transverse magnetic field to the waveguide wall. Between the oscillator sections there was a polarization-selective waveguide coupler which allows simulating the branching of the useful power to the load. Thus, the initial microwave pulse could be fed to the single-mode waveguide port 1, from which a part of the power (from 20 to 60%) in the form of an x-polarized $TE_{1,1}$ wave (with the x-component of the electric field on the system axis) came in the -z direction to the input of the left polarizer, where it was converted into a rotating wave. This circularly polarized $TE_{1,1}$ wave propagates almost without losses towards the electron beam of the amplifier, reaches the reflector at the left end of the system, where, with a coefficient of 98-100%, it is reflected in the opposite + z direction, without changing direction of rotation. Propagating further in the section of the helical-waveguide gyro-TWT along with the electron beam whose particles possess the required given transverse velocity, the rotating wave $TE_{1,1}$ in the region with the increasing amplitude of the corrugation is adiabatically transformed into the operating mode of the helical waveguide, which is amplified by an electron beam at the second cyclotron harmonic. Further, this wave is converted at the corrugation exit back into the rotating wave $TE_{1,1}$, which once again passes through the polarizer, but already in the + z direction. In this case, the circularly polarized wave is converted back to the linearly polarized one, but with the y-component of the electric field on the system axis. Such a mode (practically not perturbed by the coupler) transforms into a rotating wave in the right polarizer, passes without resonance interaction towards the beam in the absorbing section, reflects without changing the direction of rotation, and on the return path (in the -z direction) is resonantly absorbed, betraying its energy into rotational energy of electrons. Passing again through the polarizer in the -z direction, it becomes x-polarized, which will result in a part of its power (from 40 to 80%) being output through the waveguide port 2. The remainder will go to the input of the amplifier, thereby closing the feedback loop.

In accordance with the planned experiments, the frequency of 34 GHz was selected as the operating

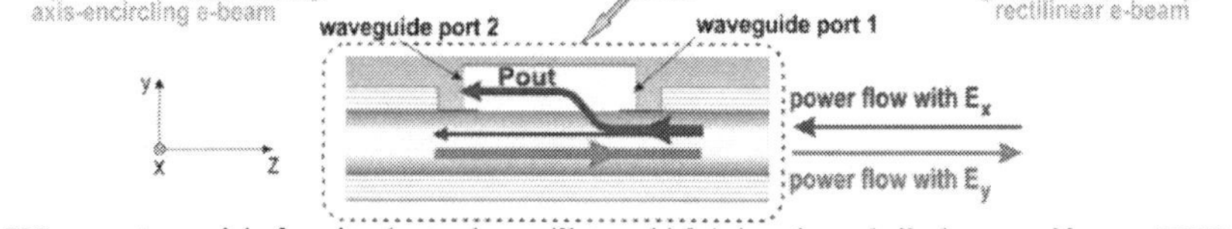

Fig. 1. 3D PIC computer model of an ultrashort pulse oscillator which is based on a helical-waveguide gyro-TWT and a resonance cyclotron absorber with a rectilinear electron beam.

one. The gyro-TWT was fed by an encircling electron beams with the following parameters: electron energies of 70 keV, beam current of 6...10 A, pitch factor 1 ... 1.3, magnetic field of 0.68..0.7 T. Parameters of corrugation were chosen in such a way as to provide an optimum amplification of ultrashort pulses [2].

At the initial stage, in order to clarify the parameters of the saturable cyclotron absorber, it was simulated separately. The aim of optimization was providing the necessary nonlinearity and the frequency bandwidth of the absorption. The most optimal properties were achieved for the following parameters of an electron beam: the energy of 50 keV and electron current of 1.2 A. The operating mode was a fundamental mode of regular round waveguide with a radius of 3 mm and the length of 83 mm. The guiding magnetic field of 1.07 T was chosen in such a way as to provide the grazing between the electron line and operating mode dispersion characteristic. Under such conditions, local interaction (i.e. interaction with minimal slippage) of every segment of the microwave pulse with a corresponding fraction of the electron beam takes place. Such arrangement is equivalent to the requirement of minimization of the absorber relaxation time in laser physics [3]. For chosen parameters the frequency dependence of the absorption ration is presented in Fig. 2 for different levels of incident power. One can see that the system exhibits properties of a saturable absorber when a relatively weak signal (P_{in} = 10 kW in Fig. 2) is attenuated significantly stronger than a signal with high power (P_{in} = 100 kW in Fig. 2). It is important that these properties are maintained over a wide frequency range.

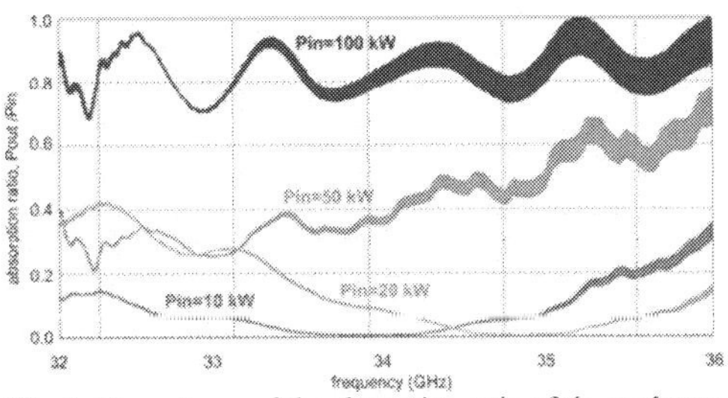

Fig. 2. Dependence of the absorption ratio of the cyclotron absorber on the frequency for various levels of incident power.

In further simulations of the complete system, the absorber and amplifier parameters are varied within certain limits in order to optimize the generation regime. The obtained results confirm the practical feasibility of the chosen concept of an USP oscillator (see Fig. 3). The highest stability of generation (pulse-to-pulse reproducibility) is achieved in the hard excitation mode when generation develops after injection of an external pulse into the system for bleaching of the absorber. In this case, the pulse repetition period is ~ 7 ns, which roughly corresponds to the delay time in the feedback loop. In the soft excitation mode, the most regular pulse sequences are obtained with a delayed switching on the absorber with respect to the amplifying section due to a linear increase in the current (Fig. 3b). It should be noted that the coherence of pulses (the time stability of the phase of microwave oscillations) is maintained throughout the counting time (in this case 300 ns), in spite of some variations in the amplitude of the generated pulses.

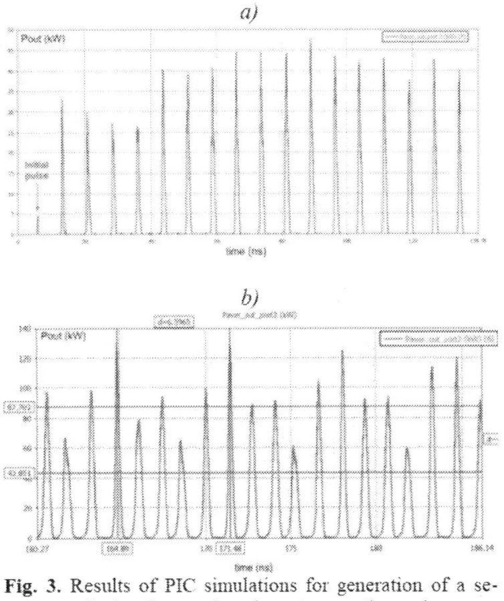

Fig. 3. Results of PIC simulations for generation of a sequence of ultrashort pulses in a two-section microwave oscillator with passive mode locking. The hard (a) and soft (b) self-excitation modes are presented.

Thus, the developed computer model, on the one hand, adequately describes the main characteristics of electron-wave interaction in the system under study, and, on the other hand, requires moderate computer resources. The simulation carried out based on this basis is a full-fledged numerical experiment that made it possible to determine the parameters for the practical realization of the ultrashort pulse generator with the passive mode-locking in microwave electronics.

This work was supported by Russian Scientific Foundation (RSCF), grant No. 16-12-01078.

References

1. *Ginzburg N.S., Denisov G.G., Vilkov M.N. et al.* Generation of "gigantic" ultra-short microwave pulses based on passive mode-locking effect in electron oscillators with saturable absorber in the feedback loop // Phys. Plasmas. 2016. V. 23. P. 050702-1-4.

2. *Ginzburg N. S., Denisov G. G., Vilkov M. N. et al.* Generation of train of ultrashort microwave pulses by two coupled helical gyro-TWTs operating in regimes of amplification and nonlinear absorption // Physics of Plasmas. 2017. V. 24. P. 023103-1-5.

3. *Haus H.A.* Mode-Locking of Lasers // IEEE J. of Select. Topics in Quant. Electron. 2000. V. 6. No. 6. P. 1173-1185.

4. *Denisov G.G., Bratman V.L., Cross A.W. et al.* Gyrotron Traveling Wave Amplifier with a Helical Interaction Waveguide // Phys. Rev. Lett. 1998. V. 81, No. 25. P. 5680-5683.

5. *https://www.cst.com*

Measuring absorptivity of ceramic materials at high temperatures in Gyrotron Ceramics Sintering System

I. I. Volkovskaya, A. Eremeev, Yu. Bykov

Institute of Applied Physics, Russian Academy of Sciences, Nizhny Novgorod, Russia, volkovskaya@appl.sci-nnov.ru

The development of processes based on the microwave heating of materials requires knowledge of the dielectric properties of the materials and their temperature dependences. In this work we measure the microwave radiation absorption coefficient in alumina-based ceramic materials over a wide range of temperatures. The method is based on comparing the Q-factors of an unloaded resonator and a resonator with an absorbing sample placed inside [1], [2]. In our experiments the working chamber of the gyrotron complex is in fact a multimode (L>>λ) untuned resonator [3]. A distinctive feature of this method is the use of radiation simultaneously for both diagnosis of properties (measurement of the absorption coefficient) and heating of the sample.

In the experiments we investigated absorptivity of two types of alumina ceramics materials: A0 (> 99.5% Al_2O_3) and AZ11 (88% Al_2O_3 + 11% ZrO_2 + 1% Y_2O_3). Samples of both compositions were presintered in a resistive heating furnace to a density of about 80% of the theoretical density (the porosity was near 20%). Basing on the measured absorptivity and theoretical calculations of the absorption coefficients of the samples under conditions of omnidirectional microwave irradiation and the dielectric characteristics of materials of complex composition, the loss tangents of the studied materials were determined.

The Q-factor of the resonator was measured using a microwave power meter connected to the resonator through a small cross-section aperture. The measured power P is proportional to the intensity of the electromagnetic field inside the resonator. Comparing the values of the power meter signals in cases of an unloaded resonator and a resonator loaded with a sample, it is possible to determine the ratio of the Q-factors in these two cases (the index 0 refers to the unloaded resonator)

$$\frac{Q}{Q_0} = \frac{P}{P_0}. \quad (1)$$

Using the relation (1) and determining from independent measurements the Q-factor of the unloaded resonator Q_0, we obtained the expression for the absorption coefficient α of microwave radiation in the sample

$$\alpha = \frac{\omega V}{cSQ_0}\left(\frac{P_0}{P} - 1\right), \quad (2)$$

where $\omega = 2\pi f$, f=24 GHz is the frequency of the electromagnetic field, c is the speed of light, V is the volume of the chamber, S is the sample surface area.

By processing the power meter data, we obtain an experimental dependence of the absorption coefficient on temperature. Knowing the values of the microwave radiation absorption coefficient in the sample and the real part of the dielectric permittivity of the sample, we can calculate the loss tangent.

Shown in Fig. 1 is one of the realization of the heating of A0 sample.

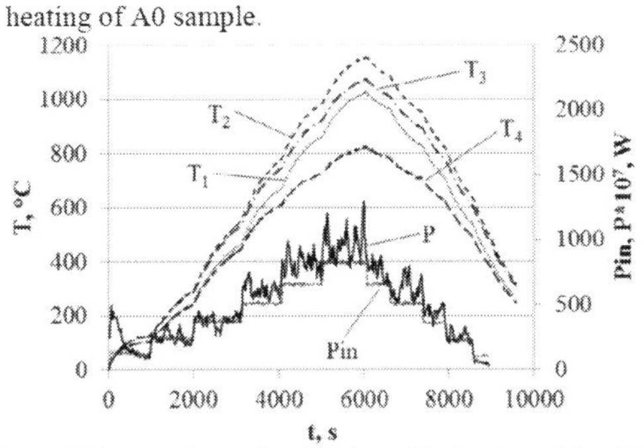

Fig. 1. The experimental realization of the heating of the A0 sample: T_1, T_2, T_3, T_4 are the temperatures at various points in the sample, P_{in} is the microwave power delivered to the working chamber, P is the power measured by the power meter

For the theoretical calculation of the absorption coefficient, we used the well-known solution to the problem of absorption of electromagnetic radiation in a plane dielectric layer of finite thickness d [2], [4]. The thickness of the samples was much smaller than their transversal dimensions, so the contribution of the side surfaces was not taken into account. The absorption coefficient α in the material was calculated according to the formula:

$$\alpha = \int_0^{\frac{\pi}{2}} A(\theta, \varepsilon', \varepsilon'', d, \omega) \sin\theta \cos\theta \, d\theta, \quad (3)$$

where $A = \frac{A_{TE}+A_{TM}}{2}$ is the average absorption coefficient, A_{TE} and A_{TM} are the absorption coefficients for TE and TM incident waves correspondingly, θ is the angle between the incident wave and the surface of the sample, ε' и ε'' are the real and imaginary parts of the effective dielectric permittivity of the sample, respectively. To calculate the effective dielectric permittivity of a sample with complicated compositions including pores, we used the effective medium approximation (Bruggeman's model) [5]. The dielectric permittivity values of Al_2O_3 and ZrO_2 at room temperature were taken from [6], [7]. In the temperature range of these experiments, the real parts of the dielectric permittivity of these oxides in the centimeter wavelength range have weak dependence on the temperature [8]. This allowed us to consider the real parts of effective dielectric permittivity as a constant value.

We can determine the temperature dependence of the loss tangent ($\tan\delta = \varepsilon''/\varepsilon'$) using the theoretical dependence $\alpha(\tan\delta)$ (3) and the experimental temperature dependence of the absorption coefficient α. The results obtained are shown in Fig. 2. Here, data on the loss tangent for alumina ceramics of similar composition (99.5 – 99.7% Al_2O_3) at a frequency of 15 GHz at room temperature [9] are given for comparison.

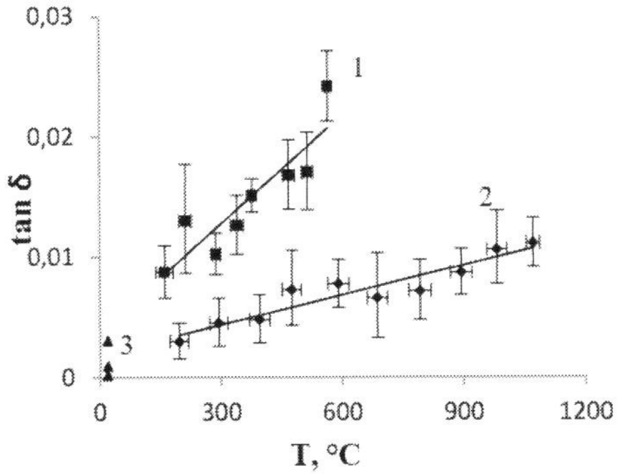

Fig. 2. Dependence of the loss tangent on temperature: 1 – AZ11, 2 – A0, 3 – alumina ceramics (99.5 - 99.7% Al_2O_3) according to [9] at f = 15 GHz

Conclusions

We obtained the temperature dependence of the loss tangent in ceramic materials of complex composition. The method is based on a comparison of the measured intensity of electromagnetic radiation in the working chamber of a gyrotron complex with and without a sample inside the chamber. The method makes it possible to carry out measurements at high material temperatures, since it does not require contact of the measuring elements with hot surfaces.

The measurements were carried out using the frequency typically used for microwave processing.

The accuracy of determining the absorption coefficients is 10-30%, which is acceptable for most problems associated with the use of microwave heating for research and development of such applied processes as sintering and joining of ceramic and composite materials.

It is shown that for known dielectric constants of the components of the measured composite materials, it is possible to determine the loss tangent as a function of the material temperature.

The following restrictions apply to the use of this method:

1) dimensions of the sample must be much greater than the wavelength of the radiation,

2) characteristic transverse dimensions of the sample must be much greater than its thickness, which is necessary for the correct use of the model of a uniformly heated plane infinite dielectric layer,

3) temperature of the sample is determined by the incident power and the exposure time, and therefore is not an independent parameter in the measurements.

This research was supported by Russian Science Foundation, grant #17-19-01530.

References

1. *Willis E. Lamb, Jr*, Theory of a Microwave Spectroscope, // Phys. Rev. 70, 308, (1946).
2. *Friedrich Kremer, Jerald R. Izatt*, Millimeter-wave absorption measurements in low-loss dielectrics using an untuned cavity resonator, // International Journal of Infrared and Millimeter Waves, p.675-683, (1981).
3. *Yu. Bykov, A. Eremeev, M, Glyavin, V. Kholoptsev, A.Luchinin, I. Plotnikov, G. Denisov, A. Bogdashev, G.Kalynova, V. Semenov, N. Zharova*, 24-84 GHz gyrotron systems for technological microwave applications // IEEE Trans. Plasma Sci., v. 32, 67–72 (2004).
4. *Jerald R. Izatt and Friedrich Kremer*, Millimeter wave measurement of both parts of the complex index of refraction using an untuned cavity resonator, // Applied Optics, Vol. 20, No. 14, p.2555-2559, (1981)
5. *David J. Bergman, David Stroud*, Physical properties of macroscopically inhomogeneous media, // Solid state physics, p.155, (1992).
6. *W.W. Ho*, High-temperature dielectric properties of polycrystalline ceramics, // Mater.Res.Soc.Symp. Proc., v. 124, p.137, (1988).
7. *J. Molla, R. Heidinger, A. Ibarra, G. Link*, Dielectric properties of alumina/zirconia composites at millimeter wavelengths // preprint CEIMAT-736, Centro de Investigaciones Energéticas, Medioambientales y Tecnológicas, Madrid, p.1-4, (1994)
8. *Arthur R. von Hippel*, Dielectric Materials and Applications // Journal of The Electrochemical Society 102(3):68C (1955)
9. *R. Vila, M Gonzalez, J. Molla, A. Ibarra*, Dielectric spectroscopy of alumina ceramics over a wide frequency range // Journal of Nuclear Materials, v. 253, p. 141–148 (1998).

https://doi.org/10.1051/epjconf/201818701023

Prediction of ohmic losses in miter bend polarizers

D. Wagner[1], F. Leuterer[1], W. Kasparek[2], C. Lechte[2], J. Stober[1] and ASDEX Upgrade

[1]Max-Planck-Institut für Plasmaphysik, Boltzmannstr.2, D-85748 Garching, Germany
[2]Institut für Grenzflächenverfahrenstechnik und Plasmatechnologie, Universität Stuttgart, Pfaffenwaldring 31, D-70569 Stuttgart, Germany

The polarization required to couple the output beam of Electron Cyclotron Resonance Heating (ECRH) systems efficiently to the plasma is usually provided by a set of two grooved polarizer mirrors [1,2]. These polarizers can convert the linear polarization of a gyrotron output beam into any arbitrary output polarization. Since the ohmic losses of these mirrors can significantly exceed the losses of plane mirrors (up to 3-4 times, depending on their groove geometry and depth), they are considered as one of the most critical elements in high power long-pulse ECRH transmission lines with respect to cooling. This is especially the case when such mirrors are incorporated into miter bends in closed evacuated corrugated waveguide transmission lines like the ones for ITER. This is due to the limited diameter of the mirror and the centrally peaked power profile of the HE_{11} mode.

The ohmic loss of polarizer mirrors has been investigated experimentally at high power experiments using test mirrors mounted in a miter bend of an existing high-power transmission line (Fig.1). For the high-power tests several polarizer mirrors were fabricated out of stainless steel (VA 1.4311) and copper. First tests were performed using stainless steel mirrors in order to get an easily measurable high temperature difference together with a slow heat transfer towards the mirror edges. The thickness of the test mirrors was 5 mm. The polarizer mirrors have a sinusoidal corrugation with depth 0.8 mm and 0.57 mm respectively, and a corrugation period of 1.07 mm. For comparison a plane disk, also with a thickness of 5 mm of the same material, was prepared. The mirror diameter is 140 mm covering the 87 mm inner diameter of the corrugated waveguides. All mirrors under test were equipped with a PT-100 platinum resistance temperature detector at the center of the back side whose dimension is 2 x 2 mm, thus averaging over 2 corrugation periods. The sensitivity is 6°K/V. At the end of each AUG transmission line the beam is directed into a short-pulse calorimetric load via another miter bend with directional coupler. For the polarizer tests one of the gyrotrons was operated at 140 GHz into such a calorimetric load. For the tests we applied an output power of 375 kW with a pulse length of 50 ms resulting in easily measureable temperature differences of several degrees. The response time of the PT-100 detectors on the back side is about 0.5 s. Fig.2 shows a typical temperature signal following a 50 ms gyrotron pulse. The maximum is reached 4 s after the pulse start. For the central position of the sensor there is only a very small radial temperature gradient, and therefore the radial heat transfer is small compared to the heat transfer to the backside, as can be seen in the slow decay of the temperature signal after reaching the maximum compared to the fast rise of the temperature right after the 50 ms gyrotron pulse. The maximum temperature can thus be taken as a measure of the energy deposited on the front mirror surface by the beam. After each pulse the mirror was cooled back to room temperature by pressurized air. Using the polarizer mirrors in the matching optics unit any required input polarization onto the miter bend mirror under test can be set

A strategy to minimize the losses for given mirror geometries has been developed [3]. For the usually applied polarizer systems consisting of a polarization rotator (groove depth : ~$\lambda_0/4$, with λ_0 being the free space wavelength) together with an elliptical polarizer (groove depth :~$\lambda_0/8$) the combination of polarizer angles exhibiting the lowest losses can be selected of four possible solutions. Fig.3 shows a comparison between measured and calculated losses of the four solutions for a polarization rotator. For other polarizer systems the number of solutions can be less [4]. The losses in Fig.4 are given by a normalized absorption coefficient a_{norm}, normalized to the absorption of a plane mirror with perpendicular incidence. Additionally the signal from a directional coupler at the end of the transmission line is plotted which remains approximately constant, thus verifying the same output polarization for all four solutions.

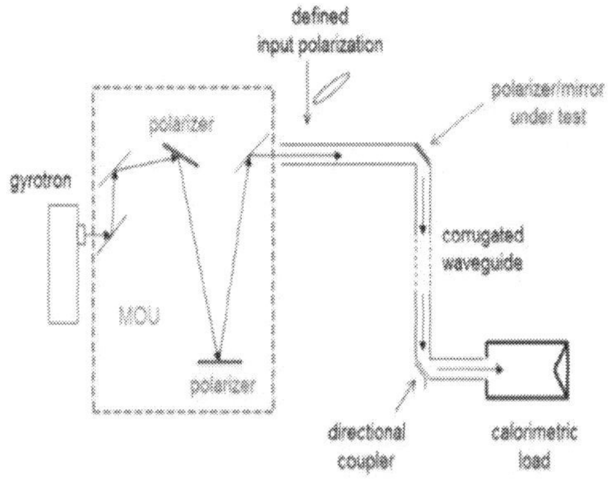

Fig. 1. Measurement setup.

 https://doi.org/10.1051/epjconf/201818701023

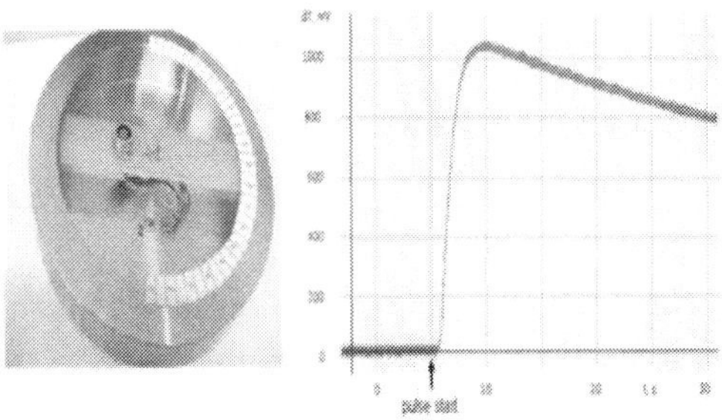

Fig. 2. Test mirror with PT-100 detectors (left) and measured temperature sequence of the central PT-100 detector (right).

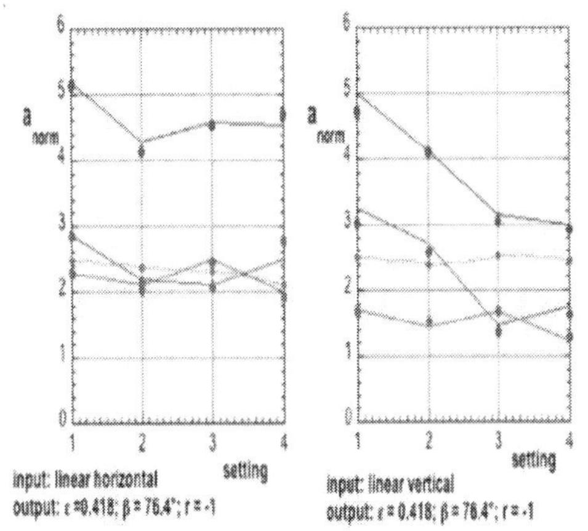

Fig.3 Measured and calculated normalized losses for the four possible settings leading to an elliptical output polarization. Left: horizontal linear input; right: vertical linear input. Dots: experimental data; lines: model calculation

References

1. *J.L. Doane et al.*, Polarizer miter bends for high-power microwave transmission: Ohmic loss and cooling, Fusion Engineering and Design, Vol. 102 (2016), 99-107.

2. *D. Wagner, F. Leuterer*, Broadband polarizers for high power multi-frequency ECRH systems, Journal of Infrared, Millimeter and Terahertz Waves, Vol. 26 (2005), 163-172.

3. *D. Wagner et al.*, Minimization of the ohmic loss of grooved polarizer mirrors in high-power ECRH systems, Journal of Infrared, Millimeter and Terahertz Waves, Vol. 37 (2016), 45-54.

4. *F. Leuterer et al.*, Experimental study of Ohmic losses of polarizer mirror system. Journal: EPJ Web of Conferences, 2017, Volume 149, Page 03002, DOI: 10.1051/epjconf/201714903002.

Problems And Solutions For Collector Systems Of Powerful Gyrotrons

V.E.Zapevalov

Institute of Applied Physics, Nizhny Novgorod, Russia, zapev@appl-sci-nnov.ru

Serious problem of high-power gyrotrons is high level of the energy of the spent electron beam after generation of microwave radiation. In a typical high-power continuous mode gyrotron with an electron beam power of 3 MW after generation of 1 MW microwave radiation (which corresponds to 33% efficiency), the power of 2 MW will be concentrated in the spent electron beam, which should be scattered by collector [1-4]. Without the use of special actions to o increase the current landing collection area, the power density at the collector is excessive and leads to its rapid damage. The use of CPD collectors somewhat improves the situation [1-4]. Typical efficiency gyrotron without CPD is 30-35%, and with CPD - about 50%. In addition to a very noticeable increase in efficiency, the use of a CPD collector greatly reduces the power dissipated on the collector. Thus, with an increase in efficiency from 33% to 50%, in a megawatt gyrotron, the power at the collector is halved. The collector is the most massive, large-sized and very expensive subsystem of powerful gyrotrons. Nevertheless, the collector is still one of the most frequently damaged parts of the gyrotron. In addition to the collector itself, the collection system includes a magnetic system for transporting the electron flow, a HV power supply system, a cooling system that provide intense cooling water flows (about 1 m^3 / s per 1 MW of power), an overload protection system, ets.

A review of variants of collector systems from the point of view of electron optics is given in [5]. Later we will dwell on the simplest systems that are close to practical implementation or are already in use at the present time. In most powerful gyrotrons with continuous or long-pulse operation collectors with longitudinal scanning of the electron beam are used [6]. Such a gyrotron as an electronic device is shown in FIG. 2, contains an electron gun and an electro-dynamic system (not shown in the diagram), a collector - 1 with a current-receiving surface -2 in the form of a symmetrical body, a vacuum case - 8 (corpus), a magnetic system - 9 for transporting of an electron beam, and an additional solenoid 3, fed by a periodically varying current. The electron beam 4 in the region of the collector 1 with the minimum field of the additional solenoid moves along the trajectory 5, with the negative field of the additional solenoid along trajectory 6, and for the maximum value of the field along trajectory 7. Thus, the electron beam 4 scans due to the action of the alternating magnetic field of the additional solenoid 3 on the current-receiving surface. As a result, the scanning zone of the electron beam 4 can significantly exceed the zone of the beam settling in the absence of the alternating field of the additional solenoid 3. Reduction of the heat flux density on the current receiving surface of the collector and, thus, the temperature of this surface is achieved by distributing the electron beam over a larger surface area of the collector with a time-varying magnetic field produced by an additional solenoid 3 surrounding the collector; a current receiving surface is profiled. The analog of this technical solution is the electron microwave device [7], where the uniformity of the heat flow density along the surface of the collector is achieved in this case by profiling the thickness of the winding of an additional solenoid, or fast transversal field [9].

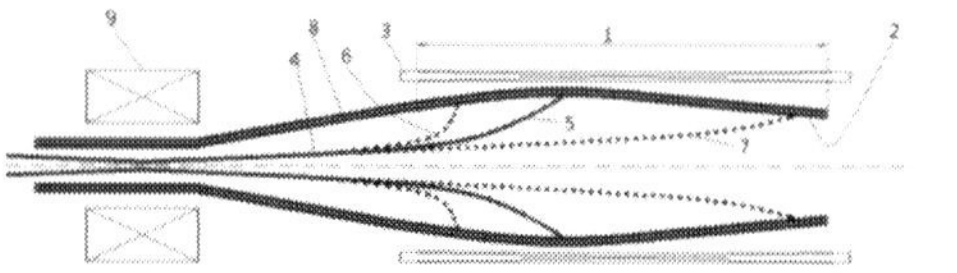

Fig. 1. A typical scheme of a powerful gyrotron with a collector, using longitudinal scanning.

The qualitative dependence of the temperature of an arbitrary point of the current-sensing collector surface in the steady-state continuous mode on time is shown in Fig.2. It can be seen that this method has a significant disadvantage: the maximum temperature of the current receiving surface is significantly higher than that could be obtained with a static distribution of the heat flux over the entire settling zone, the scanning principle itself implies a periodic effect of an electron beam on an arbitrary point of current-sensing surface of the collector and, consequently, the presence of temperature fluctuations of this point in time with the scanning frequency. The range of these oscillations obviously decreases with increasing scanning frequency (see Fig. 2).

But the scanning frequency is limited because of the screening of the alternating magnetic field by the conducting collector. Vacuum casing is made of a material with high electrical conductivity; the collector of the device is simultaneously a part of the vacuum casing. In practice, the scanning frequency is not more than a few Hertz, which is clearly not enough to eliminate temperature fluctuations.

For the body of such a collector, close in shape to a long thin nonmagnetic cylindrical shell, the characteristic frequency at the beginning of a significant screening is [8]:

$$f_{shield} = \frac{1}{\pi \sigma \mu_0 R h}, \qquad (1)$$

where, $\mu_0 = 4\pi \cdot 10^{-7}\ Гн/м$, σ - is the conductivity of the material of the collector 1, in $Ohm^{-1}m$, R is the radius of the collector 1, m, h is the thickness of the collector 1, m.

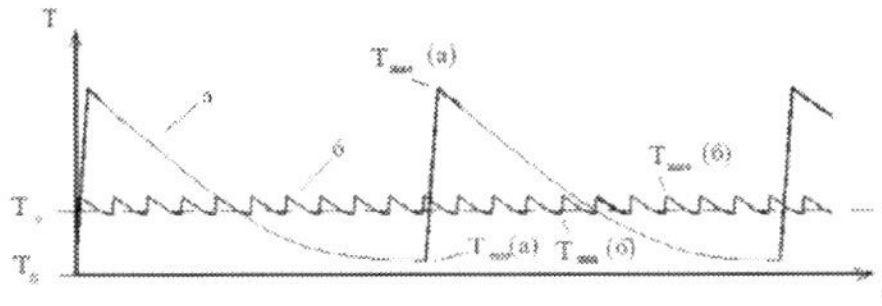

Fig. 2. The change in the temperature of the section of the collector surface versus time at a slow and fast scanning frequency

In [10], a collector is proposed in the form of an isolated single or multi-spiral coil made of a pipe through which a cooling liquid flows. For a helix in which closed circular currents are impossible, the characteristic circular frequency of the beginning of a significant screening coincides with the skin-effect frequency

$$f_{skin} = \frac{1}{\pi\sigma\mu_0 h^2} \qquad (2)$$

Thus, this solution allows, for example, to increase the scanning frequency by a factor of $\frac{f_{skin}}{f_{shield}} = \frac{R}{h}$ compared to the above described collector (all other things being equal). For a typical gyrotron copper collector (R_{coll} = 10 cm and thickness h = 5 mm), the frequency gain will be 20 times. Thus, the use of a spiral gives a noticeable gain in the allowable scanning frequency. This will significantly reduce the maximum temperature of the surface of the collector at the same power of the spent beam, and significantly increase the longevity of the instrument or significantly increase the permissible value of the power of the spent electron beam with the same longevity of the device.

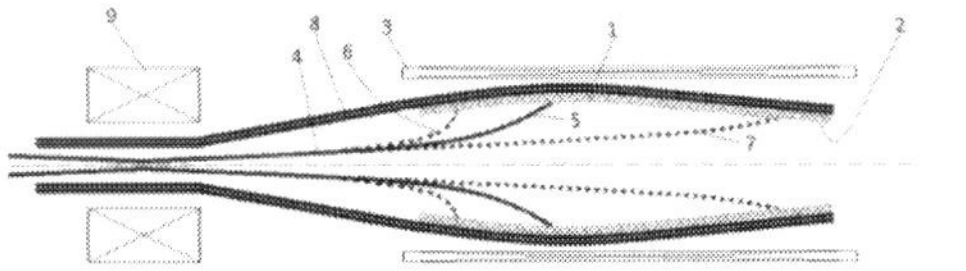

Fig. 3. Scheme of a powerful gyrotron with a spiral collector, using longitudinal scanning.

Another option is the use of a collector in which an external surface of the symmetrical body is used for current-landing (see Fig. 4) [11]. In this case, the housing of the electronic device can be made of a material with a low electrical conductivity (for example stainless steel) and the shielding is significantly weakened.

$$f_{corp} = \frac{1}{\pi\sigma_{corp}\mu_0 R_{corp} h_{corp}}, \qquad (3)$$

where σ_{corp} is the conductivity of the material of the vacuum body 1 of the microwave device, R_{corp} - its radius in *m*, and h_{corp} - its thickness in *m*. Thus, it is possible to increase the scanning frequency by a factor of $\frac{f_{corp}}{f_{coll}} = \frac{\sigma_{coll} R_{coll} h_{coll}}{\sigma_{corp} R_{corp} h_{corp}}$ compared to the variant Fig. 1 (all other conditions being equal).

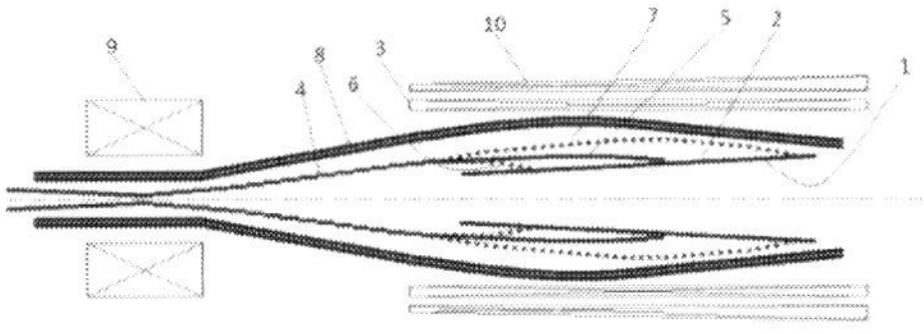

Fig. 4. The scheme of a powerful gyrotron with a collector inside the electron beam, using longitudinal scanning.

For example, for a typical powerful gyrotron, a collector made of copper (σ_{coll}= $60 \cdot 10^6$ *ohm⁻¹m*) has a radius of 100 *mm* and a thickness of 5 *mm*. Let the device have the same internal collector. For example, for R_{coll} = 110*mm* (material stainless steel 12X18H10T (σ_{corp}= $1{,}4 \cdot 10^6$ ohm^{-1}m)), and h_{corp} = 2 mm. The gain in the maximum allowable scanning frequency will be almost two orders of magnitude.

References

1. *A.V. Gaponov, et al.*, Author's certificate No. 223931 "A device for generating electromagnetic oscillations of the centimeter, millimeter, and submillimeter wavelength ranges."

2. *G.G.Denisov, et al.* Development in Russia of high-power gyrotrons for fusion. // Nuclear Fusion, 48, №5, 2008, 5 pp.1-5.

3 *M. Thumm*, State-of-the-Art of High Power Gyro-Devices and Free Electron Masers, Update 2015 (KIT Scientific Reports: 7717), 2016. https://doi.org/10.5445/KSP/1000055471

4 *A.G. Litvak et al.*, Development in Russia of megawatt power gyrotrons for fusion, // J. Infrared Millim. Terahertz Waves 32 (3) (2011) 337-342, https://doi.org/10.1007/s10762-010-9743-8.

5. *V.N. Manuilov, et al,* Gyrotron collector systems: Types and capabilities // Infrared Physics and Technology, 91 (2018) 46-54. doi:10.1016/j.infrared.2018.03.024

6. *Yu D. Larichev, A. Sh Fiks*, The Microwave Device, Author's Certificate of USSR, Patent SU 1238617 A, USA.

7. *F.Faillon, G.Mourier et al.* Elecron collector for electron tubes, U S Patent Number 4,933,594 Date June 12, 1990.

8.*V.V. Vasiliev, et al.* Conductive shells in a pulsed electromagnetic field. M., // Energoatomizdat, 1982, p.879 .

9. *V.Erckmann, G.Dammertz, M.Schmid.* Patent EP2150965A1 - Method and apparatus for collector sweeping control of an electron beam 2010-02-10

10. *Zapevalov V.E., Fix A.Sh.* IAP RAS, Electron microwave device, Patent No 2576391, Application: 2014146356/07, 18.11.2014, (priority 18 November 2014), Published: 10.03.2016 Bul. № 7

11. *Zapevalov V.E., Fix A.Sh*, IAP RAS, Electronic microwave device, Patent No. 2630251 Application No. 2016113110, priority date 05/04/2016. The date of state registration in the State Register on September 6, 2017.

Frequency Tunable sub-THz Gyrotrons for Spectroscopy Applications

A. E. Fedotov[1], I. V. Zotova[1], M. Yu. Glyavin[1], R. M. Rozental[1], A. S. Zuev[1], N. S. Ginzburg[1], A. S. Sergeev[1], and T. Idehara[2]

[1]Institute of Applied Physics of the Russian Academy of Sciences, Nizhny Novgorod, Russia, zotova@appl.sci-nnov.ru
[2]Research Center for Development of Far-Infrared Region, University of Fukui, Japan

Development of methods of sub-THz medium-power (0.01 to 1 kW) radiation generation with wide-band (several per cent) frequency tuning provides new possibilities for relevant fundamental and engineering applications. In particular, direct measurement of positronium hyperfine structure [1] would allow an experimental verification of a number of quantum electrodynamics predictions. On the other hand, CW tunable coherent radiation with specified parameters can be used in spectroscopy for detection of very small-intensity molecular spectra using the method of radio-acoustic detection of absorption (RAD spectroscopy [2-4]). Most powerful and simultaneously most compact and accessible devices operating in the sub-THz range are the gyrotrons which have already confirmed their significance for a number of present-day issues. At the same time, one of gyrotron features is fairly narrow (less than percent) frequency tuning band which is a substantial limitation for gyrotron applications.

The main obstacle for implementation of a gyrotron with wide frequency tuning band is high Q-factor of the operating mode with one longitudinal variation near the cutoff frequency. Thus, for provision of smooth frequency tuning, excitation of higher longitudinal modes with relatively low Q-factor in the long resonators was usually used [5,6]. However, in this case, the part of Ohmic losses raise dramatically and, on the other hand, sensibility to the velocity spread in the electron beam increases. Both reasons lead to significant decrease in output power and strongly limit the frequency tuning.

In this paper, in contrast to most of the previous works, we propose using a relatively short cavity with low Q-factors to provide efficient excitation of a number of high-order axial modes. The main idea of the proposed method is exploiting the weaker sensitivity of a short-cavity gyrotron to the velocity spread in the electron beam. In fact, on a short resonator length the velocity spread does not have time for a significant influence on the electron-wave interaction process. At the same time, it is obvious that the shortening of the resonator will require increasing the current of the gyrotron electron beam. In addition to providing the starting conditions for exciting high-order longitudinal modes, it will also provide overlapping of the generation zones at neighboring modes. At the same time, the required current value can be substantially reduced by going to operation at low transverse modes due to the growth of the electron-wave coupling coefficient. These considerations make it possible to expect that, with a certain optimization, such approaches can provide a sufficiently wide band of gyrotron generation.

Further using the KARAT code [7] we carried out PIC simulations for two gyrotron schemes. The first gyrotron with a frequency of about 200 GHz, a power of 0.5-1 kW and a tuning band of about 5% is intended for direct spectroscopy of positronium. The second gyrotron with a frequency of about 164 GHz, a power of 0.1-0.3 kW and a tuning band of about 3% is planned to be used for RAD spectroscopy experiments. The main parameters for both schemes are shown in Table 1.

Table 1. Parameters of tunable sub-THz gyrotrons

Operating voltage	20 kV	15 kV
Beam current	1.5-2A	0.4A
Pitch-factor	1.0	
Operating mode	TE_{12}	TE_{02}
Cavity radius	1.25 mm	2 mm
Cavity length	12-16 mm	20 mm
Beam injection radius	0.72 mm	0.55 mm
Operating frequency	~204 GHz	164 GHz
Tuning band	~5%	~3%
Output power	0.5-1 kW	0.1-0.3 kW

Further results of PIC-simulations are presented in detail for the case of 200 GHz gyrotron. The geometry of interaction space and instantaneous positions of charged macroparticles are shown in Fig.1a,b. The simulations evidence that for the velocity spread of 20% the radiation power exceeds 0.5 kW in the entire frequency tuning band of 10 GHz (5%) (see curve 1 in Fig.2). For this region of the magnetic field, from 7.3 to 8.6 T, stationary single-mode excitation of the operating TE_{12} mode takes place (Fig. 1c). For a magnetic field higher than 8.6 T, the excitation of the parasitic transverse TE_{51} mode occurs, which limits the further increase in the bandwidth. The use of the proposed gyrotron operation regime with a short cavity and a high electron current ensure a wide frequency tuning band even for a very high spread, 50%. This result confirms the small influence of the velocity spread on the electron-wave interaction for short-cavity gyrotron.

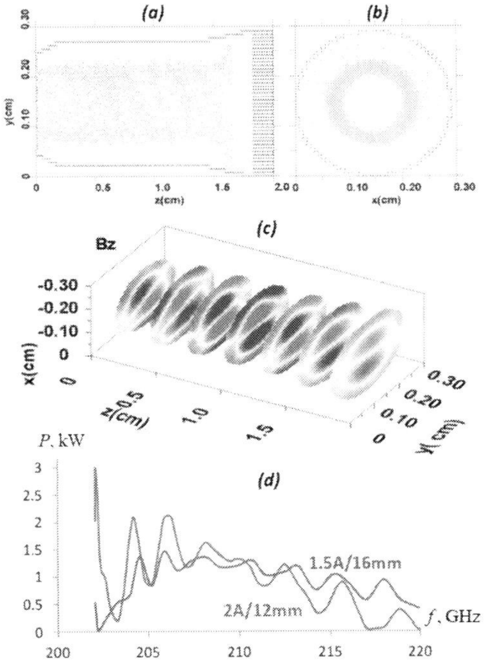

Fig.1. (a, b) Geometry of the gyrotron cavity used in 3D PIC simulations of 200 GHz gyrotron. (c) The transverse structure of the excited TE_{12} mode. (d) The gyrotron output power vs. radiation frequency for different cavity lengthes and electron currents

It should be noted that for the further improvement of the output characteristics of a tunable gyrotron, methods for reducing ohmic losses and eliminating gaps on the power vs frequency curve should be employed. These possible methods, which are the subjects of the further research, include a smooth taper at the output end of the cavity [6,8], a cavity profile taper [9], and a cathode-end power output configuration of the device [10,11].

The work was supported by the Russian Science Foundation under grant No. 18-12-00394.

References

1. *Yamazaki T., Miyazaki A., Suehara T., Namba T., Asai S., Kobayashi T., Saito H., Ogawa I., Idehara T., Sabchevski S.* Direct Observation of the Hyperfine Transition of Ground-State Positronium // Phys. Rev. Lett. 2012. V. 108. No. 25. Art.no. 253401.

2. *Belov S.P., Kazakov V.P., Krupnov A.F., Markov V.N., Mel'nikov A.A., Skvortsov V.A., Tret'yakov M.Yu.* The study of microwave pressure lineshifts // J. Mol. Spectros. 1982. V.94. No. 2. P. 264-282.

3. *Li H., Le Roy R.J.* Quadrupole moment function and absolute infrared quadrupolar intensities for N_2 // J. Chem. Phys. 2007. V. 126. No. 22. Art.no. 224301.

4. *Ba Y.A, Wenger C., Surleau R., Boudon V., Rotger M., Daumont L., Bonhommeau D.A., Tyuterev V.G., Dubernet M.-L.* MeCaSDa and ECaSDa: Methane and ethene calculated spectroscopic databases for the virtual atomic and molecular data centre // J. Quant. Spectrosc. Radiat. Transf. 2013. V.130. P.62-68.

5. *Chang T.H., Idehara T., Ogawa I., Agusu L., and Kobayashi S.* Frequency tunable gyrotron using backward-wave components // J. Appl. Phys. 2009. V. 105. No. 6. Art. no. 063304.

6. *Barnes A.B., Nanni E.A., Hertzfeld J., Griffin R.G., and Temkin R.J.* A 250 GHz gyrotron with a 3 GHz tuning bandwidth for dynamic nuclear polarization // J. Magn. Resonance. 2012. V. 221. P. 147-153.

7. *Tarakanov V.P.* Code KARAT in simulations of power microwave sources including Cherenkov plasma devices, vircators, orotron, E-field sensor, calorimeter etc. // EPJ Web of Conferences. 2017. V. 149. Art.no. 04024.

8. *Torrezan A.C., Shapiro M.A., Sirigiri J.R., Temkin R.J., Griffin R.G.* Operation of a Continuously Frequency-Tunable Second-Harmonic CW 330-GHz Gyrotron for Dynamic Nuclear Polarization // IEEE Trans. Electron. Dev. 2011. V. 58. No. 8. P. 2777-2783.

9. *Qi X.-B., Du C.-H., Pan S., Ji X., Huang B., Liu P.-K.* Terahertz Broadband-Tunable Minigyrotron With a Pulse Magnet // IEEE Trans. Electron. Dev. 2017. V.64. No. 2. P. 527-535.

10. *Chang T.H., Chen S.H.* Stepwise frequency tuning of a gyrotron backward-wave oscillator // Phys. Plasmas. 2004. V. 12. No. 1. Art.no. 013104.

11. *Bratman V.L., Fedotov A.E., Kalynov Yu.K., Osharin I.V., Zavolsky N.A.* Smooth Wideband Frequency Tuning in Low-Voltage Gyrotron With Cathode-End Power Output // IEEE Trans. Electron Dev. 2017. V. 64. No. 12. P. 5147-5150.

Waveguide DC breaks for TE_{01} and HE_{11} microwave transmission lines

D. I. Sobolev, A. P. Gashturi, G. G. Denisov, G. I. Kalynova, D. A. Lukovnikov

Institute of Applied Physics, Nizhny Novgorod, Russia, nrtnm@mail.ru

Introduction

Oversized waveguide are widely used in high power and high frequency microwave setups. Low modes in such waveguides have relatively low losses in gaps several wavelengths long [1], which make it possible to use waveguide gaps for DC insulation. Further optimization of DC breaks is possible by introducing waveguide taper converters at both sides, which are designed to increase beam quality and increase waveguide radius at its ends. The method of synthesis of such converters developed earlier by authors allowed to decrease the losses by an order of magnitude or more in comparison to simple break [2], but improvements are possible to make the converters smaller and keep losses near the theoretical minimum limited by diffraction.

Waveguide system synthesis with multi-region field analysis

The optimized DC break consists of two waveguide converters separated with air or vacuum gap and some other components, which may be exposed to high-power electromagnetic field, e.g. insulated box around the gap and some parts of cooling system. Direct analysis of fields in the whole system may take quite a long time, and synthesis usually requires several tens of calculations. To increase the speed, we separated the calculations into several parts (Fig. 1). Fields in region II can be calculated by fast methods like coupled modes or mode-matching, but region III of arbitrary shape requires slow direct methods. In low-loss DC break, only a few low modes are present inside the waveguide converters, therefore we can calculate the radiation of each of these mode from transmitting waveguide cross-section into region III once and then decompose the results into several waveguide modes at the receiving end to form a low-dimension transmission matrix. In general case, this approach has very low precision, but in a special case where transmission efficiency is very high, it produces good results and allows us to skip slow direct field calculations within the iteration loop.

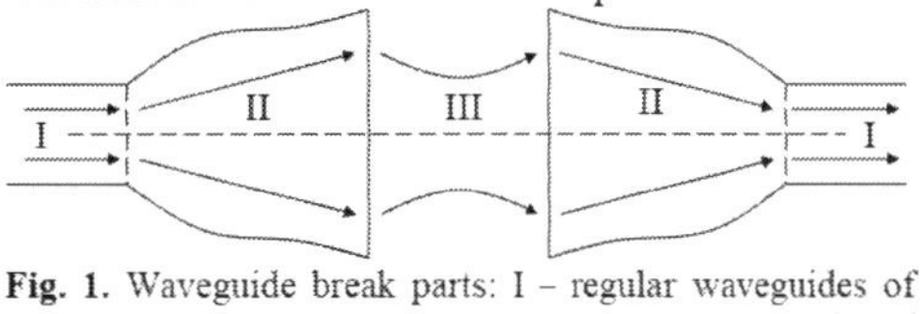

Fig. 1. Waveguide break parts: I – regular waveguides of transmission line, II – waveguide converters, III – interior of insulated box.

The synthesis of converters is based on the previous work made by authors [3]. At the beginning of every iteration step the system (Fig. 2) consists of the given input cross-section S_1, given output cross-section S_2 and initial conducting wall surface S_0. If we consider only one orientation, then conversion efficiency can be is increased after iteration by calculating two field distributions inside the waveguide. The first one is calculated using the given $\mathbf{E}_1$ and $\mathbf{H}_1$ fields at the input as a boundary condition, the second one by using the spatially reversed desired $\mathbf{E}_2$ and $\mathbf{H}_2$ fields at the output. Small and smooth deformation l orthogonal to waveguide wall affects the efficiency (mutual power) P according to formulas

$$\Delta P = \int_{S_0} lF\,dS, \quad F = -ik\left(\mathbf{H}_{\tau 1}\mathbf{H}_{\tau -2} + \mathbf{E}_{n1}\mathbf{E}_{n-2}\right) \quad (1)$$

where τ and n means tangential and normal field components, respectively. We can minimize the average square of deformation (i.e. norm) using real and imaginary parts of (1) as equality constraints with given ΔP, and in this simple case the deformation takes the form

$$l = \alpha \operatorname{Re} F + \beta \operatorname{Im} F \quad (2)$$

where constants α and β are expressed by ΔP and surface integrals of function F [4].

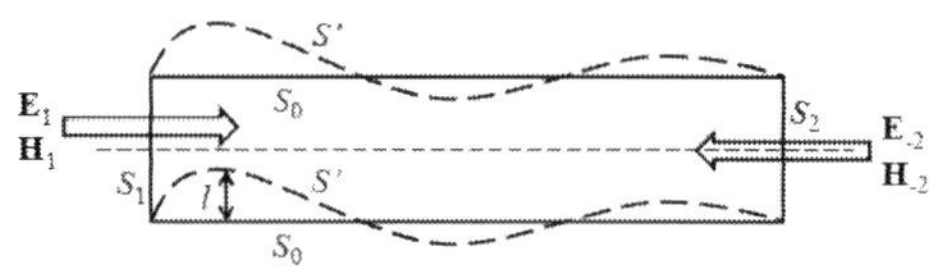

Fig. 2. Waveguide synthesis notations: two given waveguide cross-sections S_1 and S_2 (input and output, respectively), the waveguide surface before iteration S_0 and the surface after iteration S', the deformation value l at some point of the surface, the specified input fields $\mathbf{E}_1$ and $\mathbf{H}_1$, and the reversed desired output fields $\mathbf{E}_{-2}$ and $\mathbf{H}_{-2}$.

At the first iteration the fields in the simplified representation are far from the real ones, but as the synthesis converges the accuracy increases.

DC break for TE_{01} mode

The break for TE_{01} transmission line have to be able to withstand 80 kV voltage, the operating frequency of the setup is 28 GHz, and the radius of the transmission line is 16.27 mm. The gap length equal to 100 mm is enough withstand the required voltage with allowed static field strength 30 kV/cm, as the field bigger near the waveguide ends. The waveguide radius at its ends have to be 50 mm to transmit the wave beam with very low losses. The length of each synthesized waveguide converter is 318 mm, and the total calculated efficiency is 99.8% at the central frequency (Fig. 3). The DC break passed the tests at voltages up to 85 kV limited by the voltage source available, and low power microwave tests which show the transmission efficiency 99%±1%.

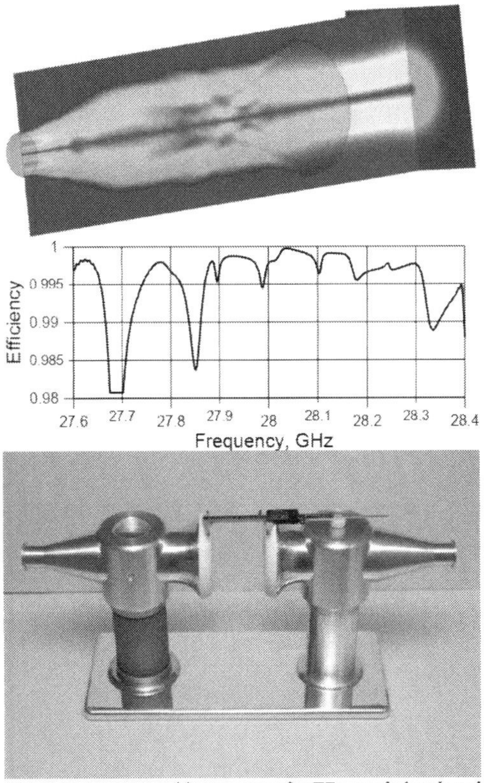

Fig. 3. Top – waveguide converter for TE_{01} mode break and cross-section of the electric field; middle – calculated transmission efficiency (FDTD); bottom – DC break in configuration used for high voltage tests.

Wideband DC break for HE_{11} mode

Special wideband version of DC break was synthesized and tested for HE_{11} mode. Able to withstand 50 kV, this break has waveguide converters 270 mm length each with radius changing from 20 mm to 31.75 mm, gap length 63.5 mm and was designed to work in frequency band 28.5-31.5 GHz. Calculated efficiency is at least 97.5% in 5% frequency band and at least 95% in 10% frequency band. Measured efficiency agrees well with calculations in 5% band, but drops in 10% band (Fig. 4), which may be the result of the limited precision of low-dimension transmission matrix.

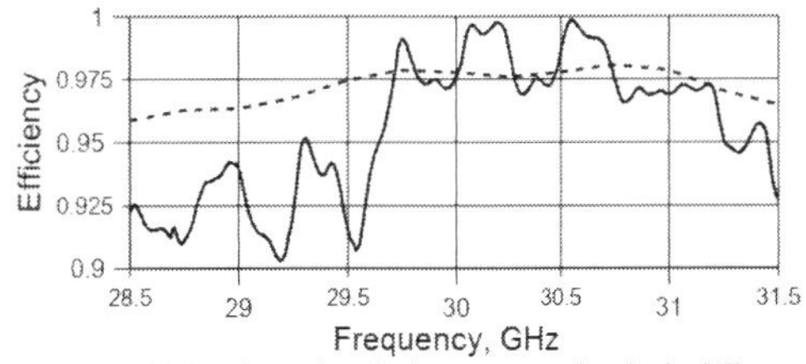

Fig. 4. Wideband DC break for HE_{11} mode: dashed line – calculated transmission efficiency, solid line – measured transmission efficiency.

References

1. *Wagner, D., Thumm, M., Kasparek, W., Müller, G. A., and Braz, O.* Prediction of TE-, TM-, and hybrid mode transmission losses in gaps of oversized waveguides using a scattering matrix code // Int. J. Infrared and Millimeter Waves. 1996. V. 17. No. 6. P. 1071-1081.

2. *Sobolev, D.I., Chirkov, A.V., Denisov, G.G., Lukovnikov, D.A., Malygin, V.I.* Minimization of diffraction losses in big gaps of multi-mode waveguides // Int. J. Infrared and Millimeter Waves. 2005. V. 26. No. 7. P. 953-966.

3. *Sobolev, D. I., Denisov, G. G.* Principles of Synthesis of Multimode Waveguide Units // IEEE Trans. Plasma Sci. 2010. V. 38. No. 10. P. 2825 – 2830.

Contributions to the Joint DFG-RSF Project - Generation of Ultra-Short Microwave Pulses -

A. Marek[1], K. A. Avramidis[1], S. M. Copplestone[2], N. S. Ginzburg[3], S. Illy[1], J. Jelonnek[1,4], J. Jin[1], S. V. Mishakin[3], P. Ortwein[2] and M. Thumm[1,4]

[1]IHM, [4]IHE, Karlsruhe Institute of Technology (KIT), Karlsruhe, Germany, alexander.marek@kit.edu
[2]Institute of Aerodynamics and Gas Dynamics, University of Stuttgart, Stuttgart, Germany
[3]Institute of Applied Physics, Russian Academy of Sciences (IAP- RAS), Nizhny Novgorod, Russia

Introduction

In a joined DFG-RSF project of the Institute of Applied Physics (IAP-RAS) and the Institute for Pulsed Power and Microwave Technology (IHM-KIT), the generation of a periodic sequence of powerful short RF pulses is studied [1]. Such powerful pulses in the millimeter and sub-millimeter wavelength range can be useful for a large number of fundamental problems and practical applications, including diagnostics of plasmas, photochemistry, biophysics, new locating systems, and spectroscopy of various media.

A periodic sequence of powerful pulses can be generated in a feedback loop consisting of an amplifier and a saturable absorber [1]. The saturable absorber acts as a nonlinear filter which transmits high intensity signals while signals with low intensity are absorbed. In such a feedback loop, the periodic signal is generated by the mechanism of passive mode-locking. In the millimeter wavelength range, gyro-devices as the gyro-traveling-wave-tube (gyro-TWT) are well suited for the realization of both devices, the amplifier and the nonlinear absorber.

Components of a feedback loop

Amplifier and absorber can be realized as gyro-TWTs with helical interaction region [2]. These devices provide a series of advantages compared to classical gyro-TWTs based on cylindrical interaction circuits with dielectric losses. The helical gyro-TWTs can operate at the 2th cyclotron harmonic and therefore, the required magnetic field is reduced by the factor of 2 compared to classical gyro-TWTs operating at the fundamental cyclotron harmonic. Further advantages are that helical gyro-TWTs offer a broad frequency bandwidth up to 20%, which is important for the generation of ultra-short pulses in the sub-nanosecond range, and that they are not sensitive to electron velocity spread. Also, it is possible to operate helical them with a single window for the input and output of the RF signal [2]. In the single input/output window operation it is possible to couple even high-power signals into the device, which is an essential condition in the feedback loop. In addition, the single input/output window simplifies the coupling of the amplifier and the absorber in a feedback loop.

Amplifier and absorber will operate in different operating regimes. The amplifier will run in a regime optimal for the maximal amplification of ultra-short pulses [3], while the absorber will run in the so-called Kompfner dip regime [4].

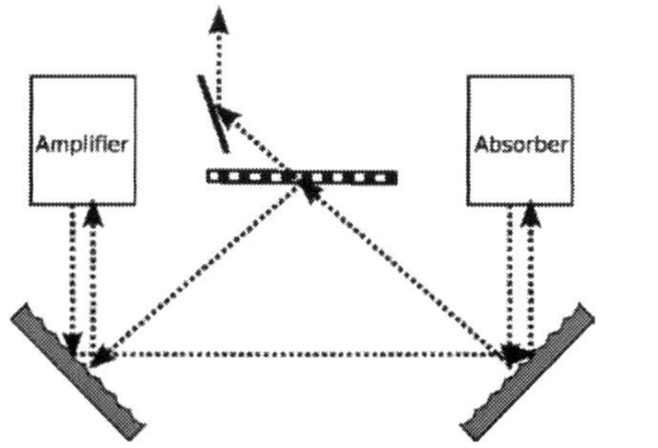

Fig. 1. Amplifier and absorber devices working with single input-output window coupled with a quasi-optical mirror system.

For an optimum amplification of ultra-short pulses, the group velocity in the helical section of the amplifier is higher than the axial velocity of the electron beam. Therefore, the RF pulse "slips" over the electron beam and accumulates energy from different fractions of the beam [3].

The absorber is designed similar to a standard helical broadband amplifier, i.e. the group velocity and the axial velocity of the electron beam should be equal over a broad frequency bandwidth. The Kompfner dip regime then is realized by a detuning of the magnetic field and the electromagnetic wave is attenuated by the electron beam.

In an experimental setup, the coupling of the two gyro-devices can be realized by a quasi-optical mirror system. To decouple a fraction of the signal oscillating in the feedback loop, a semi-transparent mirror is required. Such a semi-transparent mirror can be realized as polarizing beam-splitter based on a stack of rectangular metal plates as described in [5]. For optimum operation, the decoupling should take place only on the signal trail from the absorber back to the amplifier. Therefore, a separation of the signal trails amplifier to absorber and absorber to amplifier is necessary. For this purpose, we can take advantage of an attribute of helical gyro-TWTs in the single window operation: The output signal is cross-polarized to the input signal. As result, mirrors with sinusoidal grating can be used as effective polarization splitter. A possible quasi-optical coupling system based on the previous described technics is shown in figure 1.

Simulations

Full-wave PIC simulations including the interaction between electrons and electromagnetic waves are

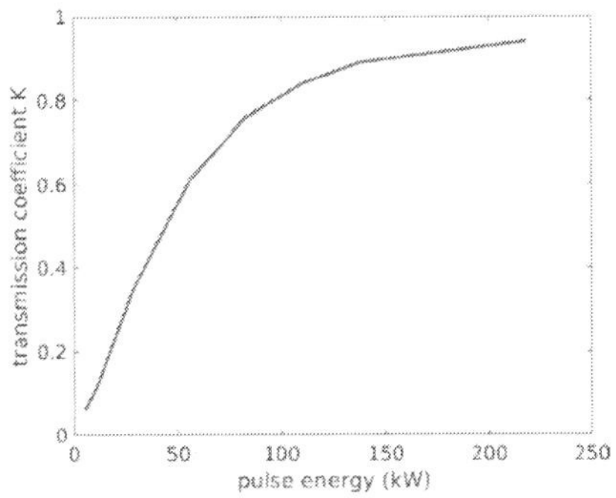

Fig. 2. Dependence of the transmission coefficient on the power of the incident signal for a helical gyro-TWT operating in the Kompfner dip regime.

performed with the advanced simulation program PICLas [6], developed by the Institute of Aerodynamics and Gas Dynamics (IAG) at the University of Stuttgart. The required high-order hexahedral meshes are generated with the mesh-generator HOPR [7], also developed by the IAG.

In the following, simulations of an absorber as suggested in [8] for a feedback loop at a center frequency of 30 GHz are performed. The suggested feedback loop is designed to provide pulses with a peak-power up to 400 kW and a width of only 0.25 ns by a separation of 5-10 ns [8]. The absorber has a mean waveguide radius of 0.4 cm with a corrugation of amplitude 0.1 cm and a period of 1.0 cm. The electron beam has a beam voltage of 54 kV, a current of 2.4 A and a pitch factor of 0.53.

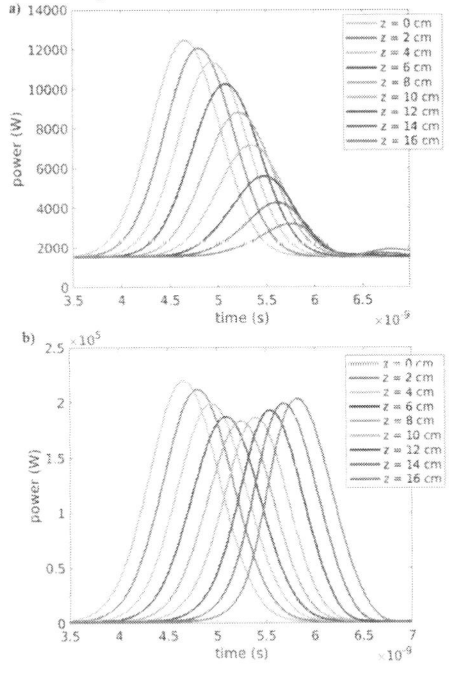

Fig. 3. RF-power of a Gaussian pulse propagating along the interaction region of a helical gyro-TWT operating in the Kompfner dip regime, for a low power input signal a), and a high power signal b).

In figure 2, the simulated transmission coefficient for short Gaussian pulses with different peak powers is shown. While weak signals are almost completely absorbed, high power signals are only weakly attenuated. Therefore, the simulations proof the qualification of gyro-TWTs operating in the Kompfner dip regime as saturable absorbers. From figure 3, the saturable absorber effect can be understood in detail. In figure 3 a), the propagation of a 10 kW Gaussian RF pulse along the absorber is shown. The absorption of almost the complete pulse can be clearly seen. In figure 3 b), the propagation of a high-power input pulse (200 kW) along the same absorber is shown. It can be seen that first a maximum fraction of 30 kW of the input power is absorbed by the electron beam. The transfer of the RF power to the kinetic energy of the electrons causes a shift from the Kompfner dip regime to the amplification regime. Thus, the RF pulse gains a fraction of its previous emitted energy back. A comparison of the input and output signal shows that the high-power pulse lost only a small fraction of its power. Therefore, the device exhibits properties of a saturable absorber.

Acknowledgements

The research is supported by the joint RSF-DFG project (Je 711/1-1, 16-42-01078) "Generation of Ultrashort Pulses in Millimeter and Submillimeter Bands for Spectroscopy and Diagnostic of Various Media Based on Passive Mode-locking in Electronic Devices with Nonlinear Cyclotron Absorber in the Feedback Loop".

References

1. Ginzburg, N. S., et al. "Generation of a periodic sequence of powerful ultrashort pulses in a traveling wave tube with bleachable absorber in the feedback loop." *Technical Physics Letters* 41.9 (2015): 836-839.
2. Denisov, Gregory G., et al. "Microwave system for feeding and extracting power to and from a gyrotron traveling-wave tube through one window." *IEEE Electron Device Letters* 35.7 (2014): 789-791.
3. Ginzburg, N. S., et al. "Mechanisms of amplification of ultrashort electromagnetic pulses in gyrotron traveling wave tube with helically corrugated waveguide." *Physics of Plasmas* 22.11 (2015): 113111.
4. Ginzburg, N. S., et al. "Generation of trains of ultrashort microwave pulses by two coupled helical gyro-TWTs operating in regimes of amplification and nonlinear absorption." *Physics of Plasmas* 24.2 (2017): 023103.
5. Mendis, Rajind, et al. "Artificial dielectric polarizing-beamsplitter and isolator for the terahertz region." *Scientific Reports* 7.1 (2017): 5909.
6. Munz, Claus-Dieter, et al. "Coupled particle-in-cell and direct simulation Monte Carlo method for simulating reactive plasma flows." *Comptes Rendus Mécanique* 342.10-11 (2014): 662-670.
7. Hindenlang, Florian, Thomas Bolemann, and C-D. Munz. "Mesh curving techniques for high order discontinuous Galerkin simulations." *IDIHOM: Industrialization of high-order methods-a top-down approach*. Springer, Cham, 2015. 133-152.
8. Ginzburg, Naum S., et al. "Ultrawideband Millimeter-Wave Oscillators Based on Two Coupled Gyro-TWTs With Helical Waveguide." *IEEE Transactions on Electron Devices* 65.6 (2018): 2334-2339.

https://doi.org/10.1051/epjconf/201818701028

Current Status of the KIT Coaxial-Cavity Long-Pulse Gyrotron and its Key Components

S. Ruess[1,2], K. A. Avramidis[1], G. Gantenbein[1], Z. Ioannidis[1], S. Illy[1], P. C. Kalaria[1], T. Kobarg[1], I. Gr. Pagonakis[1], T. Ruess[1], T. Rzesnicki[1], M. Thumm[1,2], J. Weggen[1] and J. Jelonnek[1,2]

[1]IHM, [2]IHE, Karlsruhe Institute of Technology (KIT), 76131 Karlsruhe, Germany, Sebastian.Ruess@kit.edu

Abstract

Today's performance requirements for future DEMO gyrotrons are an operating frequency > 200 GHz and an output power of 2 MW. A total gyrotron efficiency of better than 60 % must be achieved. Multi-purpose/multi-frequency operation and frequency step-tunability are required also. It has been shown earlier, that the coaxial-cavity technology is a promising candidate [1] for future fusion power plants. In [1] a world record RF output power at 170 GHz of 2.2 MW has been presented in short-pulse operation (~ few milliseconds). However, it has to be proven that the coaxial-cavity technology can be used in steady-state operation. That shall be achieved by upgrading the existing 2 MW 170 GHz short-pulse prototype with a cooling system and the implementation of advanced Magnetron Injection Guns (MIGs) [3]. Both, the advanced MIGs as well as the long-pulse gyrotron will show the way towards a possible solution for DEMO with a significantly higher output power at higher operating frequencies.

Bake-out Procedure

After the significant modification, the gyrotron was baked-out in the in-house oven up to 300 °C. All the components have been designed for a bake-out procedure of more than 400 °C. Especially the solder connections withstand a local temperature of more than 600 °C. In order to record the temperature, 12 thermocouples are connected to the mirror box. The complete bake out procedure has been controlled by a special control system to which a scenario has been programmed. The bake-out scheme consists of three major phases. In the first phase the temperature in the oven is stepwise increased up to the nominal bake out temperature of 300 °C. In the present case, 7 days were spent for the first phase. In the second phase the temperature is kept constant at the nominal value of bake-out for further 7 days. Subsequently, in the third phase, the heating of the oven is switched off and the temperature is decreasing slowly in order to avoid stresses in the components due to different thermal expansions.

The usually occurring event is the increase of vacuum pressure during the ramp-up process. If the pressure exceeds a value of more than p1 = 1.0×10^{-6} mbar the temperature is kept constant until the pressure becomes less than p2 = 5.0×10^{-7} mbar (marked in Fig. 1 with event 1). If the pressure is smaller than p2 the oven increases the temperature again. If the pressure is exceeding the pressure level p3=5.0×10^{-5} mbar the control stops immediately the heating process and shuts down the oven in a controlled manner (marked in Fig. 1 with event 2).

During the bake-out process the vacuum pressure could be reduced from 2×10^{-7} mbar down to 1×10^{-8} mbar, which is the lower limit of the vacuum turbo pump.

Alignment of the Coaxial Cavity Insert

The effective gyrotron operation requires an exact alignment of the mechanical gyrotron axis. I.e. the axis of the electron gun, the axis of the resonator and the axis of the insert with respect to the axis of the magnetic field. Therefore the alignment of the gyrotron axis relative to the axis of the magnetic field has to be satisfied in order to guarantee a contentric electron beam with respect to the cavity and insert. Due to the very tight tolerance requirements it is very important to measure the accuracy of the mentioned gyrotron components at the nominal operation condition, and, if necessary to correct them. The measurement of the electron beam position relative to the insert and cavity wall can be done by the help of the dipole coils of the Oxford Instruments gyrotron magnet at KIT. These coils offers the possibility to shift the electron beam in the xy plane.

In order to measure the position of the hollow electron beam relative to the insert the compression of the magnetic field has to be increased. In this case the electron beam radius is decreasing and moving

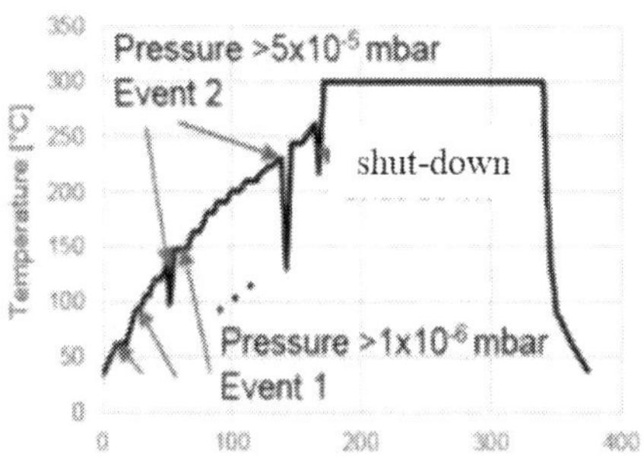

Fig. 1: Temperature profile during tube bake-out.

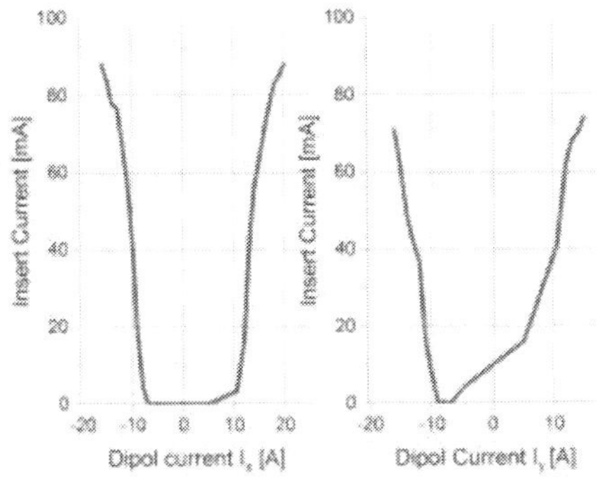

Fig. 2: Measurement of the insert current at different excitations of the dipole currents in xy-plane.

closer to the insert. By the use of the dipole coils the electron beam is touching the insert locally. The current to the insert, measured versus the dipole current, is shown in Fig. 2. In order to prevent damages at the insert, the measurement was done with reduced electron beam parameters. The beam current was 0.335 A with an acceleration voltage of 1 kV. The measurement of the insert current versus the dipol currents in x- and y- directions are shown in Fig. 2. Based on the results the deviation between the axis of the electron beam and the axis of the insert is in x –direction 0.036 mm, while the displacement in y-direction is 0.032 mm. The relation between dipole current and shift of the electron beam Δrbeam is given by: Δrbeam = 0.012IX,Y/A [mm].

The misalignment, mentioned above, of the insert is negligible and therefore a stable gyrotron operation can be expected.

Fig. 3: The coaxial-cavity long-pulse gyrotron in the Oxford Instruments superconducting magnet with the calorimeter for the RF power measurement.

Experimental Verification of the Long-Pulse Coaxial Cavity Gyrotron

After the bake-out procedure and conditioning of the gyrotron the completely in-house manufactured and assembled components have to be verified and compared with the pre-prototype 2 MW 170 GHz short-pulse coaxial-cavity gyrotron. For a better comparison, it was decided to use the configuration of the new long-pulse gyrotron with the already existing MIG of the pre-prototype short-pulse gyrotron. This has the advantage that the performance uncertainty of the MIG can be excluded and the focus can be concentrated to the cavity and RF beam forming components.

In Fig. 3 the coaxial-cavity long-pulse gyrotron [3] is shown in the superconducting magnet (SCM). The measurements and analysis of the RF signal has shown that the gyrotron is operating absolutely stable at a single frequency and therefore in only one cavity mode. The measured mode series is in agreement with the calculated prediction.

At ~85 kV a mode switch from the nominal mode $TE_{34,19}$ at 169.86 GHz to the neighboring mode $TE_{33,19}$ at 167.79 GHz takes place. The maximum measured output power at the nominal mode is 2.1 MW with an efficiency of ~30 %, which is in conformance with the simulation results. With increasing magnetic compression the pitch factor is increasing and therefore the transverse kinetic energy of the electrons and the RF output power are also increasing. This effect was experimentally verified and the maximum measured RF output power is 2.2 MW with an efficiency of 33 %. A further RF power increase is not limited by the gyrotron technology itself. The main limitation is the capability of the SCM and power supply. The new longer-pulse gyrotron has a similar performance as the short-pulse pre-prototype gyrotron, however the efficiency is 3 % higher.

Acknowledgements

Part of this work is carried out within the framework of the EUROfusion Consortium and is receiving funding from the Euratom research and training programme 2014-2018 under grant agreement No 633053. The views and opinions expressed herein do not necessarily reflect those of the European Commission.

References

1. T. Rzesnicki, et al., IEEE Trans. Plasma Science, Vol. 38, No. 6, pp. 1141-1149, 2010.
2. S. Ruess, et al., „An Inverse Magnetron Injection Gun for the KIT 2-MW Coaxial-Cavity Gyrotron", IEEE Transactions on Electron Devices, (63), p. 2104-2109, 2016, DOI:10.1109/TED.2016.2540298.
3. S. Ruess, et al., "KIT coaxial gyrotron development: From ITER toward DEMO", International Journal of Microwave and Wireless Technologies, 1-9. doi:10.1017/S1759078718000144

AUTHOR INDEX

AUTHOR INDEX